GARDEN
OF
MICROBIAL DELIGHTS

A · PRACTICAL · GUIDE · TO · THE · SUBVISIBLE · WORLD

GARDEN OF MICROBIAL DELIGHTS

A · PRACTICAL · GUIDE · TO · THE · SUBVISIBLE · WORLD

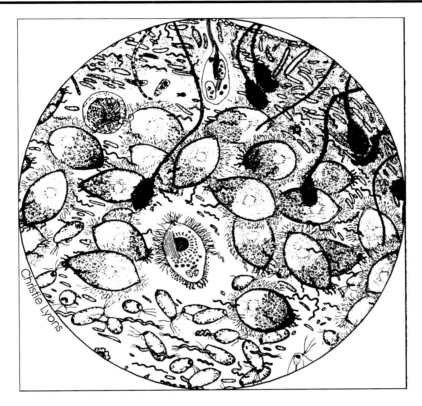

DORION SAGAN
LYNN MARGULIS

KENDALL/HUNT PUBLISHING COMPANY
2460 Kerper Boulevard P.O. Box 539 Dubuque, Iowa 52004-0539

CONTENTS

DESCENT INTO THE MICROCOSM

O·N·E
■ THE REALM OF THE MICROBES ■

T·W·O
■ OUR ULTIMATE ANCESTRY ■

T·H·R·E·E
■ BIOSPHERE AND BIOTA ■

F·O·U·R
■ MAKING A LIVING AND STAYING ALIVE ■

F·I·V·E
▪ GROWTH, MATURATION, AND SEX ▪

S·I·X
▪ MICROBIAL INTERACTIONS ▪

III

USING THE FIELD GUIDE

S·E·V·E·N
▪ NAMES ▪

E·I·G·H·T
▪ COLLECTING AND KEEPING YOUR MICROBE ▪

III

THE GUIDE TO THE MICROCOSM

T·W·E·L·V·E
▪ FUNGI ▪

T·H·I·R·T·E·E·N
▪ THE MACROCOSM AND BEYOND ▪

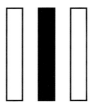

DESCENT · INTO · THE · MICROCOSM

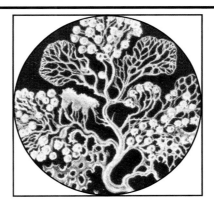

ANTONII a LEEUWENHOEK,

Regiæ Societatis ANGLICANÆ Socii,

ARCANA
NATURÆ
DETECTA.

Editio Novissima, Auctior & Correctior.

LVGDVNI BATAVORVM,
Apud JOH: ARNOLD: LANGERAK,,
M D C C XXII.

ANTONII a LEEUWENHOEK. 41

exiguus horum fere semper erat numerus. Secundum genus simile erat Fig. B. hæc sæpe turbinis in modum circumagebantur, ac aliquando instituebant cursum, ut in C. & D. ostenditur; hæc multo majori erant numero. Tertii generis figuram dignoscere non potui, nam aliquando videbantur esse figuræ oblongæ, aliquando perfecte rotundæ; hæc adeo erant exigua, ut majora non apparerent, quam Fig. E. ac præterea tam celeriter progrediebantur, ut per se hinc & inde ferrentur, atque ac si magnum, culicum aut muscarum sine ordine volitantium numerum videremus. Hæc ultima mihi quidem ita apparuerunt, ut putarem, me videre aliquot millia in aliqua aquæ parte, vel salivæ cum supra dicta materia permixtæ, quæ aliquo arenæ grano non erat major; licet ibi quidem partes novem essent aquæ vel salivæ, & una saltem materiæ, quam ex dentibus incisoriis sive molaribus detraxeram. Porro constabat maxima pars materiæ ex immensa striarum multitudine; quarum quidem una ab alia longitudine plurimum differebat, unius tamen ejusdemque erant crassitiei, aliæ incurvatæ, aliæ rectæ, ut in hac Fig. F. quæ sine ordine jacebant & quia antehac animalcula eandem habentia figuram vidi in aqua viventia, idcirco omni molimine contendi, ut observarem, utrum in illis esse vita: sed nullum motum, ex quo minimum vitæ conjiceres, potui animadvertere.

Cepi quoque salivam ex binarum fœminarum ore, quas quotidie os suum colluere, mihi constat; hanc autem, quantum potæ, accurate observavi, sed nec in ea animalcula dignoscere potui; sed postea eam immiscui materiæ, etiam acu ex dentium earum interstitiis exemtæ, atque tum in ea etiam tot animalcula viva una cum particulis seu striis oblongis, de quibus supra, detexi. Contemplavi quoque salivam pueri annos circiter octo nati, neque in ea animalcula viva detegere potui: & postea etiam eam immiscui materiæ, ex dentium interstitiis hujus pueri a me exemtæ, ac tum tantam animalculorum, aliarumque particularum copiam in ea cognovi, quantam antea jam memoravi. Deinde os meum (de industria) per tres dies non ablui, ac tum materiam, quæ parva copia dentibus meis & gingivæ adhærebat, accepi, eamque æque salivæ, & aquæ puræ pluviatili immiscui, atque in ea etiam pauca quædam viva animalcula repperi.

F

Sene

Fig: G.

dicavi ei
esse. Sa
teriam ac
cui nulla
ac etiam
quam ex
postquam
separasser
motum e
semper n
tione vid

THE REALM OF THE MICROBES

▪ A PARALLEL WORLD ▪

The microbial world is full of beauty, mystery, and nonhuman splendor. Beyond the level of resolution of the human eye there exists another world, parallel to ours and rich with life. Before the 17th century no one suspected the existence of this microscopic world. Certainly no one imagined that the microcosm predated our own visible world by eons. Today we realize that in comparison to the richness and antiquity of microbes, we dwellers of the larger realm must be considered so recent and few as to be a mere epiphenomenon—just one of many curious results of billions of years of bacterial symbioses and evolution. Indeed, strange as it may seem, the reader is just a microbe's means of making another microbe.

In the century that began with the Pilgrims sailing off to settle the New World, other pioneers began to explore the newly discovered microbial world. In 1675 Anthony van Leeuwenhoek (1632–1723), a Dutch draper with no formal scientific training and one of the first to penetrate the barrier into the unsuspected subvisible world, wrote a letter to the Royal Society, the most esteemed British scientific organization of the day (Figure 1.0). It was typical of Leeuwenhoek that he spared no detail in his account of his procedure and observations:

> *I can't forbear to tell you also, most noble sirs, that one of the back teeth in my mouth got loose again, and bothered me much in eating: so I decided to press it hard on the side with my thumb, with the idea of making the roots start out of the gum, so as to get rid of the tooth; which I succeeded in doing, for the tooth was left hanging to only a small bit of flesh, and I was able to snip it off very easily.*
>
> *The crown of this tooth was nearly all decayed, while its roots consisted of two branches; so that the very roots were uncommon hollow, and the holes in them were stuffed with a soft matter.*

FIGURE 1.0

Frontispiece diagrams and text from the Latin translation of *Arcana Naturae Deteta (Hidden Nature Revealed)*, a book by Anthony van Leeuwenhoek published by the Royal Society (London) in 1722. (Courtesy Prof. Ricardo Guerreo; book from the Botanical Museum, Central University of Barcelona, Montjuich, Barcelona, Spain.)

I took this stuff out of the hollows in the roots, and mixed it with clean rain-water, and set it before the magnifying glass so as to see if there were as many living creatures in it as I had aforetime discovered in such material: and I must confess that the whole stuff seemed to me to be alive.

Peering through his handmade microscope, the ever-curious Leeuwenhoek exclaimed that the "number of these animalcules was so extraordinarily great . . . that t'would take a thousand million of some of 'em to make up the bulk of a coarse sand-grain." Leeuwenhoek called the microbes he saw "animalcules": because of their rapid movements, he thought of them as minute animals. But these microbes were no more animals than we are giant amebas. Rather, they were independent, freely swimming cells, just as we are in reality tightly packaged collections of interdependent cells.

The microbial world, as Leeuwenhoek began to see, is inhabited by many kinds of curious and wonderful beings which, though they have profound effects on our daily lives, are generally ignored. Medical practitioners deal with the dark side of microbes: the infections, the diseases, the afflictions brought on by beings overgrowing their boundaries and roaming destructively in other organisms' territories. But there are beneficial microbes as well. Microbes have had a vast positive influence in industry, science, health, history, and the biosphere. They supply the atmosphere with oxygen and other gases, fertilize the soil, purify the water, and have influenced the development of human culture from Roman architecture (baths, sewers, and aquaducts) to minor customs of etiquette, eating, and greeting. After all, we wash our hands to prevent the spread of potentially harmful microbes.

The antiquity and diversity of microbial life is astounding. Microbes have been crucial in the development of pharmaceutical, optical, and biotechnologies. Although bacteria, protists, and fungi can be our natural enemies, they are also a lifeline to the Earth's biosphere, the network of life at the surface of the planet. These microbes are an ancient, omnipresent resource and, indeed, the lineal antecedents of human beings.

Despite their scanty size, organisms invisible to the naked eye are the dominant life forms on earth. Like adolescents embarrassed to be seen with their parents, we disparage Earth's ancestral life forms and dismiss microbes as mere germs. But primeval organisms are not only the predecessors of our form of life; they exist in our midst. As our planetary elders, they have all the advantages of seniority. Without our intervention it is their nature to grow and thrive. Although we can easily imagine environments without humans, it is difficult to imagine any terrestrial surface without an immense population of microorganisms. Human skin harbors some 100,000 microbes per square centimeter, making our bodies analogous, from a microscopic point of view, to the thickest forest of the Amazon.

While often overlooked even by scientists, microbes do far more than merely occupy space or cause disease. All of the elements crucial to global life—oxygen, nitrogen, phosphorus, sulfur, carbon—return to a usable form through the intervention of microbes. Even such toxic elements as lead and mercury are processed by microbial life. Ecology is based on the restorative decomposition of microbes and molds, acting on plants and animals after they have died to return their valuable chemical nutrients to the total living system of life on earth.

Microbes are, and have been for millions of years, the dominant (most abundant) form of life in the biosphere. Much of the oxygen animals breathe and plants employ in their photosynthetic metabolism originates as a waste gas of plankton—microbes floating in the seas. There would be no agriculture were it not for the widespread presence and fertilizing effects of blue-green and other forms of nitrogen-fixing bacteria. Our forest industry, with its products in the form of furniture, shelter, paper, and so forth, owes its productivity to the continuous growth and chemical activity of microscopic fungi known as mycorrhizae. These fungi break down phosphorus and other nutrient supplies locked up in the rocks and soil into forms usable by the trees' root systems. It is unlikely forests could even grow without such fungi, and fungal symbioses (such as those involved in mycorrhizae) probably were crucial to the formation of the first forests.

With photosynthetic organisms, such as plants, algae, and some bacteria, fungi are foremost at breaking down solid rock and turning it into fertile soil. They were crucial in the movement of life to land and they may be indispensable once again, in the future, if the deserted-looking, boulder-laden landscapes of other planets are to be transformed into luxuriant gardens.

The oil and gas we burn to meet our society's increasing energy needs also result from ancient microbial processes. They are "unearned resources" in that we are tapping into microbiological work done long before our own evolution. Life-saving antibiotics from penicillin to streptomycin are products of microbial metabolism—fungal or bacterial. Moreover, many enzymes necessary for a whole slew of industrial and medical processes originated in the microbial world. Coal deposits are built up in part from microbe bodies and actions. Certain types of microbes and microbially-dominated environments led to commercial deposits of iron, uranium, and perhaps even gold. And, today, the mining of strategically important ores depends on certain bacterial leaching activities.

Our growing understanding of how cells work, including the new techniques of genetic engineering, comes from a study of microbes. Recent studies comparing genes from different organisms strongly suggest that our cells and those of other eukaryotes (organisms whose cells have a nucleus bounded by a nuclear membrane) originated in a series of symbiotic events that involved the merging of very different sorts of bacteria.

The very founding of America and many other lands, as well as numerous other historical events, have been deeply influenced by microbes. The terrible Irish potato famine, for example, which forced thousands of people to emigrate to the United States, was caused by a funguslike protoctist—a type of eukaryotic microbe. The black plague, a disease that helped cast the entire mood and intellectual climate of Europe during the Middle Ages, resulted from the spread to humans of pathogenic bacteria carried by rats. References to fungi growing along the damp walls of stone dwellings can be found in the Bible. Throughout the ages, even when their existence was not even suspected, microbes have had a deep and lasting impact upon human culture.

With the growing focus on the potential of biotechnology, many people today are researching and using microbial resources for profit; through genetic engineering, new forms of medicine and ways to improve agricultural yields are being discovered. But microbes also hold the key to natural biocides—ecologically sound, biodegradable pesticides. Some microbes, able to metabolize and collect exotic problem chemicals like petroleum and radioactive waste, may be useful in "global hygiene," in national and international cleanup programs. And the list goes on. Methanogenic bacteria already can be grown to produce the natural gas methane as an energy source for industry.

In the future, once the process of photosynthesis—which evolved first in microbes—is fully unraveled, it may be possible to produce pure hydrogen at room temperature. Sending hydrogen gas through pipelines could revolutionize human technology, ushering in a new era simply by applying the ancient technology of photosynthetic cells. By industrially copying elegant technological processes already in use (and what is more, miniaturized) within cells, people could be provided with vast reserves of energy more cheaply, effectively, and safely than ever before. By assembling machines that mimic the respiration of oxygen-using bacteria, we could quietly transform hydrogen piped through tubes into electricity. The gas could also be used for the industrial manufacture of plastics, rubber, record albums, food, and other items currently made from petroleum products. All this would be an enlargement and elaboration upon processes already taking place inside a single plant cell—processes that evolved in bacteria many millions of years ago. It appears that these supposedly inferior, lowly, simple, and primitive life forms may still have a bit left to teach us.

In evolution it is to the benefit of organisms to worry about what threatens them or affects their survival. Historically, this has been the case with our perception of microbes: we have feared and despised all of them because of the disease-carrying tendencies of only a few. Indeed, it makes sense that we would be most concerned with their adverse effects since we have had to respond to these and not to their beneficial effects in order to secure and prolong our survival. Fear induces us to avoid or destroy that of which we are afraid. But with thousands of microbes on every square inch of skin there is no avoiding or destroying them without killing ourselves.

Fortunately, not only fear but the opposing principle of curiosity, which makes us want to examine and inspect the unknown, has been preserved in our development. With the discovery of the beneficial properties and potential uses of microbes, it is our curiosity rather than our fear that has come into the vanguard. Clearly, then, the world of normally unseen organisms—beings neither plants nor animals—represents far more than simply a nuisance we would like to be rid of with antibiotics, supermarket sprays, and other disinfectants.

Having made a quick survey of the fascinating and largely overlooked, friendly side of microbes, let us take an even closer look at some of the ways microbes have beneficially influenced humankind. We have hinted at the crucial importance of microbes during prehistoric times and throughout human history, and at the equally great impact unseen and little-known organisms may have in the future. As Bernard Dixon chronicled in his book *Magnificent Microbes* (published by Atheneum), these usually inconspicuous beings have been, are, and will be indispensable in such diverse areas as textiles, dairy products, wine making, mining, farming, geology, air quality, forestry, world food supply, ecology, soil regeneration, medicine and health, pollution control, biotechnology and genetic engineering, and space exploration and human habitation of extraterrestrial environments.

Let's begin with textiles.

Flax, especially prized during the time of the American Revolution, was a staple of life throughout the 18th century and later. The first American sailing ships—Discovery, Susan Constant, and Godspeed—that cruised up the James River in Virginia in 1607, as well as the Mayflower in Massachusetts, and, indeed, vessels from all over the world in past centuries, had elaborate sails made from the plant known as flax, *Linum usitatissumum*. The process of making linen materials, such as sails, from the flax plant depends upon two microbes, *Clostridium felsineum* and *Clostridium pectinovorum*. Early American shipbuilders would have been hard pressed to find an acceptable sail material if there had not been microbial activity for the proper flax processing.

The fibers of the flax plant were also employed to make blankets, handkerchiefs, paper, and clothing. Societies centered around its use, and eventually there emerged a highly-regarded profession of "spinsters," who spun the flax strands into these valuable goods. Flax was of such commercial importance in colonial America that a school for its making was established in Boston, and England's General Howe reported that "linen goods were much wanted by the rebels." When he prepared to evacuate Boston, Howe gave an order for all such goods to be carried away with him.

The flax plant, marked by petal-shedding blue flowers (Figure 1.1), required careful weeding until ripening in June or July. The midsummer plants were pulled up methodically by their roots and laid out to dry. Later, they were carefully tied into bundles and put into moist fields or submerged at the bottom of slowly flowing streams. The *Clostridium* bacteria, ubiquitous in soil, would begin to grow on the flax, digesting a major substance of the plant called pectin for their own food. Digestion by bacteria loosened the fibers of the flax, which were then peeled away and spun into linen or used in other products.

Other historically important small plants also became useful because of microbial processes. Among them were *Corchorus* (jute) used for burlap, furniture webbings, twine, and upholstery, and *Crotalaria* (hemp) used for cordage, fish nets, and cigarette paper.

In colonial America, and in villages throughout the world, foods were not readily available at the general store. People depended upon their own livestock, land, and produce. Of course they never realized the major role performed by microbes: the differences between chemical and microbial processes did not become clear until this century. No dairy products were produced without the aid of microorganisms. To this day, the special flavors and textures of cheeses and other true dairy products are due to the introduction (inoculation) of a variety of specific microbes.

In many communities, cheese was a staple of life for centuries. Cheesemaking starts with the curdling of milk proteins, yielding a solid mass from which excess water is drained away. Cottage and cream cheeses are made by adding bacteria such as *Leuconostoc* to pasteurized milk, and most cheeses must also be ripened by bacteria and fungi. With cheddar, the same bacteria

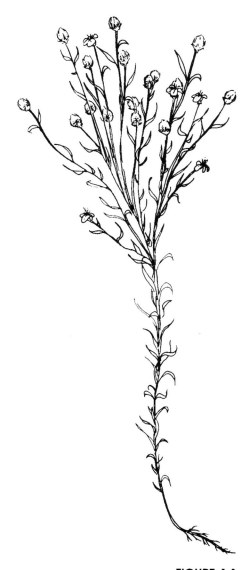

FIGURE 1.1
Linum usitatissumum, the flax plant. Fibers from this plant are used to make linen. (Drawing by J. Steven Alexander.)

7

that produce the curd also ripen the cheese. As the bacteria die, their cells release enzymes that act on the milk fat and proteins to form the many different compounds that give cheddar its particular flavor. In addition, the nutritive value of the cheese rises greatly because the bacteria also contribute an assortment of B vitamins. In the early weeks of ripening, the numbers of bacteria per gram of a given cheese reaches the hundreds of millions. High quality butter also depends upon the growth of microbes: *Leuconostoc* and *Streptococcus.*

In addition to food, alcoholic beverages are produced by subvisible life forms. Wine has been a standard drink for scores of cultures around the world for nearly 10,000 years. The Egyptians, Greeks, Romans, and others all stressed its manufacture, and its importance as a growing industry in young America was great. Nearly all alcoholic beverages, including wine, depend upon the process of fermentation. Many microbes ferment, but none except yeasts do it in ways that we can easily control. Yeasts are fungi so tiny that a microscope is required to see them clearly.

The elements of wine making are simple. Workers gather then crush and press grapes to provide the juice, or "must." Because it is both acidic and rich in sugars, must makes an ideal medium for yeast growth. To produce red wine, black grapes are mashed along with their skins and stalks. White wine is usually fashioned from either black or white grapes, but without their skins. The waxy bloom on the surface of grapes contains many microbes, including the principal one used in wine-making, *Saccharomyces cerevisiae var. ellipsoideus.* At least 150 different yeast strains are known and used to yield subtle differences in wine flavors and body. Wine makers bring the must to a fermentation vat and the resident yeasts promote vigorous fermentation, turning the sugars into alcohol and, consequently, triggering the chemistry that makes wine relatively "fruity" or "dry."

Farming presents another area in which microbes are eminently helpful rather than harmful. America possesses abundant areas of fertile, healthy soil and cropland. America's farms have been vital to her history, and the silo is necessary to any self-sufficient farm. These buildings are used to store surplus crops for the winter feedings of livestock. The fodder—whether it be sorghum, potatoes, legumes, or corn—is tightly packed into the silo with a minimum circulation of air. During storage, the indigenous microbes, numbering up to 10 billion per ounce of fodder, ferment the sugars and starches available to them from the foods. Fodder fermentation stored this way is called ensilage. Ensilage increases the nutritive value of the fodder, giving it a flavor and aroma attractive to sheep and cattle. All of this contributes to healthier livestock.

Fossil microorganisms provide us with economic benefits. Foraminifera and other marine microorganisms present in the fossil record can be found in Europe and America. The layers of soil in which they are found help indicate the geological processes that made the region. Also, pollen grains and shelled microorganisms such as radiolarians and diatoms tell us about Earth history, evolution, and oil and mineral deposits. Many of these organisms, including those known only as fossils, are extremely beautiful and intricately shaped (Figure 1.2). Because certain microfossils are always associated with oil and gas deposits, they can be used as clues and pointers to the location of underground resources.

FIGURE 1.2a

Intricately shaped skeletons of ray beings are found buried as fossils on land and as live organisms swimming in the sea. These plates from original drawings in a work by 19th century German naturalist Ernst Haeckel show various skeletal remains of ray beings, marine protists called radialarians. This plate was originally published in 1904 and reprinted by Dover Publications in 1974.

9

The oxygen we breathe every day originally was put in the air by microbes. Human beings, like other animals, depend upon the replenishment of the atmosphere's oxygen via photosynthetic organisms. Despite the vastness of forests and jungles of the world, the contribution of trees and shrubs to the Earth's overall oxygen content is less than you might expect. Microorganisms such as dinomastigotes and cyanobacteria supply the major portion of our breathable oxygen gas. More than 150 billion kilograms of oxygen are estimated to be generated by such oceanic microbes in less than a year. Oxygen bubbles are released during the normal growth of certain bacteria and all algae in the world's seas.

Microbes are crucial to life on land in other ways, for example in the development of forests. Common, characteristic trees of various regions of the world, such as the white cedar, oak, cypress, and maple, all depend upon microscopic fungi known as mycorrhizae attached to their root cells. These fungi act as conduits for soil nutrients to enter the trees' root systems; trees without mycorrhizae become stunted in their growth. Forest microorganisms also include the actinobacteria, which have recently been found to extract necessary nitrogen from the air for a wide variety of broad-leaved shrubs in the wild.

Many countries have important, agriculturally-based industries, including apple, potato, and tobacco crops. All plants need nitrogen. Yet despite the fact that it accounts for nearly 80% of our atmosphere, no plant by itself is able to obtain this important element directly from the air. Plants depend instead on actinobacteria and others present in the soil and in their roots to "fix" this atmospheric nitrogen into a usable form. Little rod-shaped bacteria called *Rhizobium* are so important in agriculture for this reason. Rhizobia swim into the plant root cells and make bulbous growths on the roots of peas, beans, and other beanlike (leguminous) plants. The rotation of crops to include legumes insures that soils are well endowed with the necessary nitrogen for high yields. Microbially-based agricultural methods have been in traditional use throughout the world. For example, cyanobacteria (blue-green bacteria) not only photosynthesize, but also fix nitrogen in the rice crops of the Far East. Without these microorganisms, rice production would drop dramatically, and starvation and nutritional losses would be even greater global problems than they already are.

When animals and plants—indeed, any organisms—die, their essential elements and compounds—nitrogen, carbon, phosphorus, and sulfur, among others—reenter the soil, water, and air, and thereby are recycled for further use. This basic principle of ecology insures that all organisms on earth will have continuing opportunities to survive and reproduce. Without an abundance of microorganisms to decompose organic matter, these cycles could not occur. Instead, dead matter would simply pile up, making the planet a huge waste dump. New life forms, and evolution itself, would be limited, if not impossible. All plants and animals depend on nutrient cycling by unseen organisms: bacteria, protoctists, and fungi. Indeed, microbes perpetually transform wastes and sludge into plant nutrients and soil so that, though individual cells live and die, the entire film of life on the Earth's surface takes on a life of its own. The ancient recycling process drives the biosphere, conferring upon it some of the traits we associate with individuality, as if it were some immense, immortal being.

FIGURE 1.2b
Soil microbes, slime molds that form spore-bearing structures as dry conditions encroach (see p. 164). The mastigote and amebal cells by which they propagate are shown at the top. These plates from original drawings in a work by 19th-century German naturalist Ernst Haeckel show various slime mold types. This plate was originally published in 1904 and reprinted by Dover Publications in 1974.

Microbes are indispensable to certain industries. There are perhaps many hundreds of microbe-mediated industrial products currently being used in the manufacture of industrial goods and services. Microbes produce as waste organic chemicals such as acetone and butanol; both are of major importance to modern industry. The anaerobic bacterium *Clostridium,* for example, is a prolific producer of these solvents, commonly used in the plastics industry. A variety of organic acids necessary in foods, pharmaceuticals, and paints come directly from microscopic fungi and bacteria. Western Europe alone produces 60 million kilograms of citric acid annually, the bulk of which comes from the fungus *Aspergillus* and its relatives as they grow, tended by human caretakers (fermentation industry employees). Itaconic and related acids are essential microbial products with many applications in the paint, adhesive, and fiber industries. Lactic acid, produced by "lactic acid" bacteria (see Chapter 10, Fermenters), is used to preserve food, to finish silk rayon fabrics, to make acrylic resins, and for a great many other purposes.

The diversity of enzymes made by microbes (but not by plants and animals) is leading to a new discipline of "enzyme engineering." Enzymes are natural large molecules that speed up chemical reactions inside living cells. Recently, for example, two enzymes associated with streptococci are proving to be invaluable. These are streptokinase and streptoDNAase. Streptokinase is used to remove clotted blood, while the DNAase directly digests DNA. Both processes are important in medical research laboratories. Contact lenses are cleaned by "enzyme treatment" with papain—a strong protein destroyer that dissolves the film of eye protein that tends to build up on the plastic lens surface. This list of examples could be greatly extended, and the field of industrial microbiology can be considered a backbone of essential goods and products in societies around the world.

Although microbes are often cited as the cause of medical problems, they are equally often the cure. Harmless viruses, pieces of viruses, or heat-killed bacteria are injected into our bloodstreams. Called "vaccines," these complex substances induce the body's immune system to produce proteins called antibodies that are destructive to the disease-causing microbe. Once produced, our antibodies stay in our bodies and destroy active viruses and bacteria that enter and circulate in the blood.

Bacteria and fungi (most famously the *Penicillium* mold, which produces penicillin) produce antibiotics—natural drugs that prevent the growth of other, more harmful bacteria and fungi. The vast majority of antibiotics now in use are derived from the selection and mass cultivation of specific strains of soil bacteria and fungi. Although many of these antibiotics can be artificially synthesized in the laboratory, microbial production by growing populations of microbes is often more efficient. Since 1945, more than 1000 different antibiotics have been isolated from fungi and bacteria. Few adults have not taken antibiotics at least once in their lives.

The study of microorganisms—particularly the mold *Neurospora crassa* and the bacterium *Escherichia coli*—has given us a tremendous body of information about the underlying chemistry of life. Modern scientific concepts of physiology, botany, and zoology depend heavily on insights that have come from microbiology. Microbiology is an integral contributor to the applied sciences, such as agriculture, medicine, and industrial biotechnology.

Even geologists and meteorologists are beginning to realize the importance of microbes for sedimentary rock formation and climate processes. In cultural and academic terms, the recognition of the microbial basis of the life of the biosphere is on a par with Newton's mechanics, Darwin's evolution, or Einstein's relativity in giving us insights into ourselves and the planet we inhabit.

And microbes are not just out there in the environment but in here, inside our bodies. Most are natural and helpful. Not only do tens of thousands of microorganisms occupy each square centimeter of the surface of our skin, but also countless others are harbored within us. There are more microorganisms present in one human being's body than there are people on Earth. Furthermore, rare individuals born without the usual capacity to carry the normal mob of skin and intestinal microbes must be carefully isolated due to their oversensitivity to infection and their weakened immune systems. Inevitably such people (with agammaglobulin anemia) die at a young age.

Even mining depends on microbes. As our natural resources diminish, we must now mine ores that were once dismissed as unworkable because of inaccessibility, cost factors, and low metal content. The search for efficient, cost-effective means to extract metals has led to the use of bacteria in order to recover copper, nickel, and even uranium. A type of bacteria found naturally in many ore deposits and waste dumps is *Thiobacillus ferrooxidans*. These microbes oxidize iron, copper, and other metal sulfides by growing in very acidic solutions. The metals are thus leached out from the sulfide rocks. Australia, Japan, Russia, Romania, Portugal, and Canada all have major microbial mining processes in operation.

Bacteria are also proving helpful to mining interests in an entirely different way—for safety purposes. Researchers at the Moscow Institute for Miners have invented a method of ridding mines of explosive methane by drilling horizontal holes in the direction of proposed future cuttings. After pumping in a liquid medium containing bacteria that thrive on methane, the methane quantity decreases to levels safe for the miners.

Utilization of microbial digestion for the removal of oil leaked by tankers into the sea is underway. Several strains of bacteria that remove oil by eating it and simultaneously converting it into edible protein have been developed by "industrial microbiologists." Tankers dump over 20 billion kilograms of oil into the world's oceans each year. We are beginning to look more and more to microbial solutions to the oil cleanup problem.

Already common in Great Britain and the United States, the use of bacteria to transform sewage waste into valuable methane gas promises to be a major energy source for the future. Sewage treatment plants operate with the help of natural technology from microbes. Ecologically, the production of methane by microbes may be vital to our survival as human beings, for methane levels in the atmosphere may act to control the amounts of the highly reactive fire-feeding gas, oxygen.

With the growing shortage of food on a worldwide scale, microbes show some promise for being efficient and highly nutritious microcrops. Photosynthetic microbes such as *Anabaena cylindrica,* a cyanobacterium, and the green alga *Chlorella* offer the greatest possibilities, since both can be easily cultivated in the laboratory. *Anabaena* is one of the best sources of vitamin B_{12}. Large algae, seaweeds, are already a major source of food in many countries: Japan and Canada have traditionally used them as supplementary food sources. Large algae require less space and less complicated production techniques than land plants in many cases.

New microbes are going to be needed for waste processing. Countless chemicals created by humans are proving to be extraordinarily troublesome to remove from the environment once they have served their purpose. Microbes capable of eating and breaking down environmental chemicals may be a partial answer to this massive problem. Some bacteria can convert certain plastics into recycling organic compounds. Bacteria such as *Cytophaga*, which digest wood products, could relieve us from our almost total dependence on landfills and dumps. Various yeasts are able to feed on pollutants, converting them into carbohydrates and proteins.

New forms of life are being found at ocean depths where extremely hot gases vent out from the Earth's crust into the sea. These vents, one of the greatest biological discoveries of the century, feature a wide variety of unusual, huge animals. Since there is no sunlight at these depths, the question immediately arose about how such large life forms—red tube worms and large clams—could exist at this level. The answer is clear—bacteria. Sulfur-metabolizing, vitamin-producing microbes of various kinds have been found in great numbers in the vent areas; most of them are living in symbiotic consortia with other bacteria or with animals. The worms depend on the chemical activities of the bacteria within them for their own nutrition and survival.

Of course, the one area of beneficial microbial action which is well known is that of genetic engineering and microbiotechnology. The transfer of portions of DNA and RNA from one microbe to another for medical, agricultural, and waste removal benefits, as well as industrial microbiology, has become a highly developed science with important ramifications. In April 1987, the first genetically altered bacterial strain, designed to make plants resistant to frost damage, was released in California. The next month another experimental strain, a genetically engineered version of *Rhizobium meliloti*, was released in Wisconsin. A symbiotic bacterium that naturally grows on the roots of peas, beans, and other plants, supplying them with nutrients, the strain raised alfalfa yields by 15% in greenhouse tests. For better or worse, with insulin and other rare medicines now being mass-produced by microbes, genetic engineering and microbio-technology offer humanity a Promethean torch of biological potential. Although barely tapped, such potential is already sending a shock wave of concern through society.

Finally, bacteria, protists, fungi, and other organisms that evolved long ago on Earth will play a key role in the space program. The next step for the Russians, United Statesians and those people of other many nations is to live for prolonged periods in space. To do this requires recycling, or biospherics. Biospherics is the new discipline concerned with maintaining life in closed environments. Miniaturized biospheres in which air, water, and food are continually recycled will be crucial in order for human beings to move into space on a permanent basis. It seems that the technology of such systems will depend heavily on the tireless subvisible travail of microbes. Creating such closed ecosystems on Earth will also give us a useful model of the biosphere, in which the effects of our actions on the environment can be studied without exposing us to harm.

Biospherics research is also useful as a last resort in conservation biology; within enclosed environments, endangered species can be protected from their polluted or otherwise disturbed natural habitats. Clair Folsome, a microbiologist at the University of Hawaii, has already kept microbes alive and well in enclosed laboratory flasks since 1967. Similarly, a firm in Arizona markets "ecospheres"—tabletop crystal spheres that contain miniature oceans that support algae and shrimp. Unlike the shrimp in traditional aquaria, the ecosphere inhabitants never have to be fed because an internal harmony of shrimp, algae, bacteria, and other microbes work to keep the nutrients cycling for years. One goal of various space programs is to set up similar environments in which the central animals are not tiny orange fairy shrimp but humans. Such ecosystems, containing not only animals for food but also people and large plants in complex environments (desert, pond, woodland), are being devised in Japan, the U.S., Russia and elsewhere. Such biospheres large enough to hold us (as well as a sufficient sample of our agricultural animals, plants, and the ubiquitous microbes needed for the waste recycling process) will be necessary if we ever reduce or sever our immediate dependence on the Earth's surface. Biospheres will also become necessary if we continue to pollute the planet so drastically that we can no longer afford to live in open environments. Ironically, severing our ancient ties with the earthly ecosystem will require the establishment in space or in sealed areas on Earth of the very microbially based ecosystems we leave behind. Although we can hypothetically get around our dependence on Mother Earth, there is no getting around the ancient microbes.

So, microbes are far more than "disease agents." The primary familiarity of the general public with the microbial world has been through historical and medical accounts of diseases and infections caused by bacteria and viruses. We have been obsessed with diseases such as tuberculosis or poliomyelitis because they can injure or kill thousands of victims each year. Certain individual diseases such as the plague *(Yersinia pestis)*, syphilis *(Treponema pallidum)*, typhoid fever *(Salmonella typhi)*, and tuberculosis and leprosy (mycobacteria) have had a devastating effect on civilization. For decades we have been ingrained with the idea of an unrelenting, good-versus-evil conflict between people and microorganisms (germs) despite the well-established fact that relatively few microbes are associated with disease compared to the number of those that are normally associated with people in good health.

The idea that microbes by themselves cause disease is simplistic. The causes of disease are far more complex than simple exposure to a "disease agent." It is really only under special conditions that microbes become dangerous, growing rapidly or producing toxins as a result of their growth. Whether or not a person becomes ill depends upon a wide array of social, cultural, and environmental conditions rather than the intrinsic characteristics of a given microbe. We are infected by potentially dangerous microbes all the time without getting sick, and some "infections," such as naturally occuring skin bacteria, are actually beneficial in maintaining good health and proper body metabolism. Skin bacteria, for instance, help prevent dangerous overgrowth by fungi such as *Candida,* also normally present on the skin. Killing bacteria can permit the growth of fungi and vice versa. What should really shock us are the millions upon millions of people who become infected with viruses and bacteria all over the world each year without suffering any harm.

Combine this with our brief overview of their positive effects and we must say that the overall balance of microbial actions tips decidedly toward the good. Although we have not paid much attention in the past to their good points, microbes have them in great abundance. Life as a complete system can work nowhere without them. As we realize their omnipresence and importance, it is time to overcome our squeamishness, temper our xenophobia with curiosity, and see if we might not use these durable and versatile beings of which we have lived so long in ignorance and mortal fear. After all, we are an evolutionary mosaic built up from them; if we are completely disgusted by them we are in a sense completely disgusting ourselves.

As we overcome such fear and disgust we find that not only direct, practical benefits but also subtler, aesthetic ones accrue from a new knowledge of microbes. These tiny living organisms produce the exotic flavors and textures of French cheeses, gourmet sauces, yogurt, fine bread, and wine. They synthesize such delightful aromas as the fresh smell of the woods and the pungent scent of the sea. Some produce dramatic effects in the human beings out of all proportion to their size. The single fungus ergot is used not only as the useful source of medicines alleviating migraines and inducing birth, but also as the precursor to the hallucinogen LSD, a substance that had a dramatic, far-reaching social and cultural impact. Other fungi, in other cultures at other times, have been employed by shamans and equally valued for their physical or psychotropic effects.

Ultimately, the powerful chemical effects on the body and brain by simple molecules produced by ancient organisms may be understood in evolutionary terms as the result of competition and cooperation among ancestral lineages. One place in which the aesthetic and practical areas of microbial life overlap is sex. There is evidence that some microbial products seem to be aphrodisiacs: certain fungi, for example, manufacture the same aromatic chemicals produced by men and women as natural sexual perfumes. If similar perfumes were developed and marketed (and they would certainly be highly profitable) along with blue jeans and wine coolers, who knows what the effects on society might be? It makes one wonder what else still lurks, half-hewn in the evolutionary dark, waiting to be formed by our further encounters with the rich microbial world.

As humans, we owe our lives to a bustling internal microbial world. Bacteria crowd our intestines, producing enzymes and vitamins. The lineal descendants of ancient bacteria even float inside the cells of all animals today. As structures called mitochondria, they provide us all with the energy we need to move a muscle or digest fat.

Plants depend upon a different long-standing partnership of cells. Inside every cell of every plant is another sort of transformed bacteria, the chlorophyll-containing plastids. Usually green, these organelles give nature its healthy verdant hue. The green of leaves, stems, and immature fruit is the green of tiny chloroplasts—once-symbiotic bacteria—that use light to make food inside plant cells.

As yet not integral parts of plant or animal cells are the trillions of microbes normally dwelling on our skin, in our bowels, and even on the surface of our eyes. As ever-present epibionts (organisms that live on the surface of other organisms), bacteria are certainly part of our bodies. Everywhere we go we take immense numbers of microbes with us; every day they reproduce on our skin, swarm over the bumps on our tongue, squiggle between our teeth and gums.

First glimpsed in the 17th century by the light (optical) microscope, the microbial world can now be revealed by far more powerful electron microscopes. With today's optical instruments we can peer into the far reaches of the microcosm. We can speculate about the private lives and ancient history of the complex cells of which we are composed. In spite of the advanced technology of the modern world, however, the light microscope still prevails. While we can discover minute cellular details with huge electron microscopes, only with the light microscope can we observe the inhabitants of the microcosm without harming them. In order to visualize any microscopic being at the very high magnifications of electron microscopy, the subject must first be killed and sliced. This need for severe treatment in preparation for microscopy is unfortunate; what arouses our curiosity most is to see live organisms. Perhaps someday an instrument will exist that we can use to study microbes at an extremely close range without first having to sacrifice them.

Advancements in microscopy have enabled us to focus on some of the enchanting details of the microcosm, but we have not yet been able to map it completely. Despite the direct impact of microbes on our lives, the microbial world largely remains as remote and mysterious as the planet Neptune or Pluto. Our knowledge of it is comparable to the rough way in which the first European voyagers knew the geography of the Americas, with their crude maps full of mistakes and empty patches. Microbes are so tiny that even our sophisticated tools can only reveal their intricacies in a coarse way. Biochemical measurements record the behavior and effects of huge populations of bacteria, never of a single bacterium. Organisms that are very different in the gases and chemicals they each produce are known only by their "crowd behavior"—their behavior as a huge group of millions of individuals.

Although no clear date for the origin of the microscope exists, a crude light microscope probably was invented first in Holland around 1590. By 1609 the great Italian scientist Galileo Galilei had heard of the Dutch invention—a tube using various lenses to magnify an object. Within six months Galileo had devised his own version of the tube. He created a telescope with a magnifying power of 32, which he used in reverse to observe insects. From a study of lenses, Galileo's German astronomer-astrologer friend Johannes Kepler described how a compound microscope could be constructed in principle. (The compound microscope uses two sets of lenses and is far more effective than the simple microscope.) But the sky-gazing Galileo and Kepler were more interested in far-off stars than the earth underfoot. Not until midway through the 17th century did detailed microscopic studies begin.

In 1660 the argumentative Englishman Robert Hooke (1635–1703) wrote his *Micrographia,* a landmark examination of the world revealed by the microscope (Figure 1.3). In *Micrographia,* Hooke illustrated how things we take for granted in the visible macrocosm are, from the microscopic view, gross distortions and illusions. He showed that snow crystals are not really uniform but distinct flakes, each one an intricate and unique hexagonal maze of crystalized water. He illustrated that the razor's smooth edge is not really straight but crooked, jagged, craggy. He showed that the dot of an "i" like the one you see here is really not circular at all but an irregular blob of ink, throwing off flares like a tiny black sun.

Hooke, one of the first to explore the microbial world, did not mount his subjects on glass slides as do modern microbiologists. He stuck them to the head of a pin. This allowed him to view them from a variety of angles and to understand the complex, overall appearance of an object. He described his original method in the introduction to *Micrographia:*

> *I endeavored (as far as I was able) first to discover the true appearance, and next, to make a plain representation of it. And therefore, I never began to make any draft before, by many examinations in several lights, and in several positions to those lights, I had discovered the true form. For it is exceedingly difficult in some objects to distinguish between a prominency and a depression, between a shadow and a black stain, or a reflection and a whiteness in the color. Besides, the transparency of most objects renders them yet much more difficult than if they were opaceous.*

Hooke made one observation that has influenced biology to the present day. He prepared a thin slice of cork, which appeared smooth to the unaided

FIGURE 1.3

Robert Hooke's microscope. An oil lamp was used for a light source before the days of electricity. The object to be studied was mounted on a pin. (Drawing by Sheila Manion-Artz.)

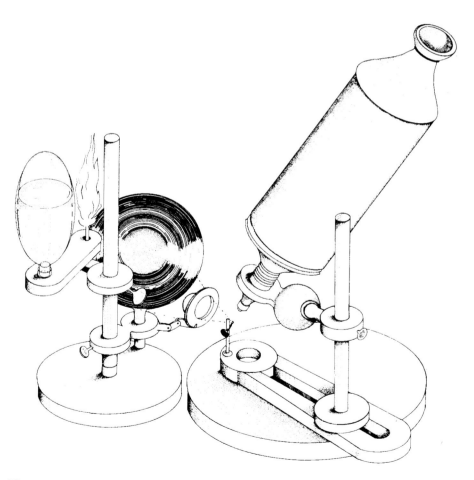

eye: "Placing it on a black object plate, because it was itself a white body . . . I could exceeding plainly perceive it to be all perforated and porous, much like a honeycomb, but that the pores of it were not regular." Hooke surmised that many small compartments of cork—the elastic outer tissue of European oak trees—also must exist in live plants. Hooke called these cubicles "cells" because they reminded him of the cuboidal chambers of the cloister. (Our word "cell" thus derives from Hooke's vision of the cells of monasteries or prisons.) After his microscopic studies, Hooke could explain why the seemingly solid cork floated in water and failed to absorb. Under the microscope, the cork cells revealed unsuspected air pockets and, therefore, the hollow cells could not absorb water. Enclosed on all sides, they trapped air like microscopic balloons, accounting for the cork's buoyancy.

Then, in the 1660s, the inveterately curious Anthony van Leeuwenhoek invented an ingenious glass-bead microscope with which he discovered his animalcules. Enchanted, Leeuwenhoek sought and found them nearly everywhere: from the gums and teeth of hard-drinking acquaintances (whose mouths, he wrongly surmised, would be too poisonous for any life form!) to pond water and rain running off rooftops. As Galileo turned his telescope skyward and discovered moons of Jupiter invisible to the naked eye, so Leeuwenhoek set his sights downward amid the plenitude of nature and discovered a different kind of new world, one that had always coexisted and overlapped with ours but had never before been seen. This parallel world is the realm of the microbes. The more it is explored, the larger it grows and the more surprises it reveals.

▪ OUR BACTERIAL ANCESTORS ▪

A look at a drop of pond water with a magnifying tube only slightly more powerful than the kind used by Galileo or Hooke will reveal that there are two sharply different kinds of life on Earth: those beings whose cells have nuclei and those beings whose cells never do. The latter, far more different from us than the little green men of science fiction, are bacteria.

As the French marine biologist Edouard Chatton first proposed in 1937, organisms that have cells with nuclei—such as animals, plants, and fungi—should be separated from those without nuclei. Chatton suggested the word "eukaryote," from the Greek words meaning *"true,"* and "nut" or "kernel," to denote all beings made of cells with true nuclei. He recognized that in contrast bacteria were "prokaryotes," or cells without nuclear membranes separating their DNA from the rest of the cell. (Pro means before, so prokaryote means cells that came before cells with nuclei.) Seaweeds, with nuclei in their cells, are eukaryotes. So are *Psiloscybe* mushrooms. So are amebas. So are we.

The gap between prokaryotes and eukaryotes is by far the greatest in all of life, and it represents a huge discontinuity in evolution. Assuming that bacteria came first, how did such tiny cells without nuclei develop into cellular giants tens of times larger, containing not only nuclei but other distinctive structures never found in prokaryotes? People, apes, bull frogs, dandelions, boars, pine trees, and truffles all have cells 10 to 100 times larger than bacterial cells. All have a central nucleus enclosed by a specialized membrane. All have distinct arrangements (chromosomes) of genetic material in their nuclei, and these contain a special kind of protein, histone, not found in bacteria.

However, cells of organisms in our visible world also contain special structures or organelles that are themselves rather like bacteria. In fact, they almost certainly are the descendants of ancient, independent bacteria that survived starvation by living and reproducing inside a larger kind of bacterial cell. These structures are the mitochondria that enliven the nucleated cell by tapping oxygen for energy, and the brightly colored plastids, which bear a striking resemblance to certain common, single-celled, photosynthetic bacteria. Trapped inside plant cells, plastids bathe in the sun, chemically using light to power the combination of air, water, and salts into stems, leaves, and flowers. Because plastids rearrange, but do not use up, the molecular components of air, water, and salts in the photosynthetic process, they must emit some molecules as waste. One plastid's waste is the oxygen upon which the mitochondria thrive. But although our cells, and those of flowers and truffles and the rest, resemble associations of certain types of bacteria, there are millions of other kinds of bacteria whose likenesses never appear inside the cells of larger beings. Compared to the manifold abilities of various kinds of bacteria to make use of and even to prompt a great diversity of chemical transformations, the cells of animals and plants (so-called "higher organisms") are extremely limited: they breathe in or respire only oxygen or carbon dioxide. Bacteria, on the other hand, can breathe sulfide, oxygen, methane, ammonia, carbon monoxide, inert nitrogen, and the like.

All the major chemical systems necessary to maintain and perpetuate life evolved in prokaryotes during the roughly two billion years before nucleated cells appeared. Some bacteria, known as anaerobes, live in the total absence of oxygen gas, while other "switch hitters" breathe oxygen when it is available and nitrates or other chemicals when it is not. Some bacteria photosynthesize—making their own food with the aid of sunlight—but in ways alien to plants. These tiny beings are purple instead of green and, instead of needing water, they require only the mephitic fumes of hot springs or the sulfur gases spewed out by volcanos or along cracks in the ocean floor. Some bacteria imbibe nitrates poisonous to us, others exhale flammable methane, and some even excrete deadly cyanide.

Bacteria also are enormously successful reproducers. To what lengths must we eukaryotic organisms go simply in order to reproduce! If a man, we must find a woman; if a female pine tree, we must wait for the yellowish pollen of the male cone to be blown past by random winds. But prokaryotes suffer no such obstacles. They reproduce whenever they want, making millions of offspring from a single progenitor in a matter of days if supplied with minimum quantities of food and water. They form cities, even civilizations, in nooks; living bacterial universes inhabit a single cranny of drying mud. Unlike eukaryotes, bacteria do not have a tightly regimented internal structure. When they duplicate themselves they do not have to mix and move their chromosomes in the slow-motion ballet of alignment and separation of eukaryotes. Prokaryotes keep their DNA afloat, jostling about within their cell bodies. And although bacteria may have thousands of times less DNA per cell than eukaryotes, they can transfer it within minutes directly to neighboring cells, sending crucial information by genetic express and creating an entirely new type of cell almost instantly in response to a changing environment. This behavior allows bacteria the world over to respond almost instantly to their local environments.

In the eukaryotic cells' favor, however, is the achievement of mitosis, the type of genetic replication and division never seen in bacteria. This intricate rite of reproduction occurs in the cells of all plants and animals. It begins with the subtle appearance and delicate spreading of protein threads, called the mitotic spindle, from the poles of the nucleated cell. Chromosomes attach to the threads and move along them to opposite ends of the cell, which then divides to form offspring cells. Bacteria have neither this nor any other kind of intracellular motion, such as the continuous cell streaming (cyclosis), which marks nucleated cells (see Figure 4.4). From the single cell of a paramecium to the millions that make up our lungs, all eukaryotic cells share this internal agility and movement.

But even the components of mitosis may have come originally from bacteria. Spiral-shaped, symbiotic bacteria called spirochetes, with their means of locomotion on the inside rather than as an external cell whip, long ago may have moved inside their hosts where they could have lost part of their anatomy and become components of our complex intracellular transportation system. By consenting to serve as various parts of our cells, these squirming bacteria may be the very foundations of our being. In any case, eukaryotes did not evolve from a vacuum, but from the earliest bacteria.

▪ THE KINDS AND KINGDOMS OF LIFE ▪

The immense difference between bacteria, or prokaryotes, and the rest of life, the eukaryotes, raises the problem of the traditional division of living beings into two kingdoms. If the microscope has revealed that the difference between bacteria and nonbacteria is the major distinction in life, what does this say about our old pigeonholing of organisms into plant and animal categories, as if there were nothing else?

As in the game Twenty Questions, everything inanimate was once considered "mineral." All living things were divided into "plant," if they were green and stayed put, or "animal," if they moved about. Except for such elusive objects as fire or shadow, the categories of plant, animal, and mineral seemed all-encompassing.

Now, however, we know that even animate things are not necessarily plant or animal. Microbes include plant–animal "hybrids," beings which are not really hybrids at all because they predate the evolution of both plants and animals. These green swimmers move like animals and photosynthesize like plants. Other curious microorganisms, such as some gas eaters (chemolithotrophs) that make their food from rocky salts and atmospheric gases, neither move nor perform photosynthesis but use inorganic chemical reactions to live and grow. Only by a dubious stretch of the imagination can such microbes be considered plants or animals. In fact, there are no such things as one-celled plants or animals. Plants and animals all grow from multicellular embryos, and many of them, such as ourselves, have cells whose numbers likely mount into the millions of billions.

21

Until the first description of the microscopic world by Anthony van Leeuwenhoek, no one guessed the extent to which the swarming world of subvisible organisms overlapped our own. True, Robert Hooke had seen fossils of microorganisms and the Italian physiologist Marcello Malpighi (1628–1694), sometimes known as the "father of microscopy," had discovered tiny capillaries containing blood. But Leeuwenhoek, originally prompted by the wish to detect imperfections in fabric arriving at his drapery shop, refined the single-lens microscope into an instrument that could magnify almost 200 times. Leeuwenhoek examined and described everything from microbes inhabiting the mouths of animals to his own swimming spermatozoa and the microscopic explosion of gunpowder, which required a special apparatus that Leeuwenhoek designed for the purpose and that nearly blinded him in the process. His British biographer, Clifford Dobell, wrote in the 1920s that Leeuwenhoek "made the maddest experiments, and attempted to see things that nobody would now even dream of seeing."

It was not until the 19th century, however, that major microscopic innovations were made, primarily in the form of the dual-lens, or compound, microscope. When properly constructed, with lenses above and below the specimen, these were immensely more powerful instruments. This basic design still used today was invented in Germany in the 1830s. These microscopes proliferated, and university professors and scholars began to describe the so-called "lower" animals and plants.

At this time, bacteria generally were defined as plants. But then many moving plantlike organisms, such as the golden-yellow alga *Ochromonas* and the euglenoids, were discovered. These were photosynthetic and often green, yet they darted through the water; some even mated like tiny marine animals. In the prevailing two-kingdom classification, such organisms were paradoxical and confusing. Nineteenth-century zoologists claimed that such organisms as euglenoids and chrysophytes were one-celled animals, while botanists, not surprisingly, presumed they were a form of plant. In retrospect this seems as silly as considering a mosquito a low type of ostrich, but at the time it was natural to compare microbes with known organisms rather than to study them in their own right.

In the late 19th century, Ernst Haeckel (1834–1919), a German naturalist and champion of Darwin's ideas of evolution, proposed a new kingdom to accommodate the taxonomically unruly microbes, which he perceived to be so fundamentally different from plants and animals as to deserve their own kingdom. At first he placed all of them in a single group, which he called the kingdom Monera. The precise boundaries of Monera changed several times during Haeckel's lifetime, alternately including and excluding nucleated organisms. But the term "moneran" didn't catch on until almost a century later, in the 1960s, and most biologists and zoologists until that time referred to bacteria, even with their alien lifestyles, as "plants."

As we shall see, however, bacteria do deserve kingdom status. Indeed, they are so deserving of separate status that calling them a kingdom may not be enough. As the Canadian scientists Sorin Sonea and Maurice Panisset put it, "Had they been discovered on Mars, their description would have been much more dramatic and the bizarre quality of their natural history, which often seems like science fiction, would not have been missed."

In 1956, the biologist H. F. Copeland of the Sacramento City College in California proposed a four-kingdom system in which he separated Haeckel's

Monera into two kingdoms, clearly distinguishing prokaryotes from eukaryotes. No one paid much attention to this idea at first. So-called "lumpers" continued to favor a two-kingdom approach while "splitters," in the minority, wanted a system having three or more kingdoms. But then, in the 1960s and 1970s, biology underwent major changes. Advances in microscopy and in the techniques of molecular biology led to recognition of the gross inadequacy of the plant-animal dichotomy. It became obvious that not only are all microbes different from plants and animals but that, even among microbes, bacteria are fundamentally different from all other forms of life on Earth.

Despite the recognition that the old taxonomy was insufficient, the scientific community was unable to make a change because of the overwhelming momentum of the old system. Not until the plant–animal split interfered with the field work of the Cornell University biologist R.H. Whittaker did something begin to be done about it. In his studies of the pine barrens of New Jersey and the deserts of the southwestern United States, Whittaker found bacteria and fungi so totally unlike plants that keeping them stuffed into the plant kingdom became an intolerable intellectual compromise. As early as 1959, he wrote an article on the broad classification of organisms in which he argued for a five-kingdom scheme of life. Instead of just plants and animals, there were now the kingdoms Animalia, Plantae, Protista, Fungi, and Monera.

Whittaker's five-kingdom classification system, with some slight modifications (such as the enlargement of Protista to include all organisms that do not develop from embryos and that are not fungi, and its subsequent renaming), serves as the basis for the discussion of the organisms detailed in this book. The Whittaker system most closely approximates the wonderful complexity of nature, and its increasing use has encouraged the study of microbes for their own sake. It rectifies the plant–animal dichotomy and vindicates Haeckel, Copeland, and other scientists who wished to dissociate themselves from the simplemindedness of an all-encompassing two-kingdom system.

The five-kingdom system denies the existence of "one-celled plants" and "animalcules." The scientific word for "single-celled animal," however, has been "protozoan," from the Greek meaning "pre-animal." This term, as the concept behind it, is misleading. If animals are defined by their development from a multicellular embryo formed by the union of egg and sperm, then there can be no one-celled animals. Even multicellular members of the protist kingdom such as seaweeds lack the tissue and organ systems we intuitively associate with animal life. For these reasons, single-celled nucleated organisms and their multicelled derivatives must be called something else, a word denoting primeval organisms without implying that they are primitive animals or plants. Whittaker used "protist" to describe members of this group and to avoid projecting animal and vegetable qualities onto types of beings some of which, judging from the fossil record, have remained essentially unchanged for more than two billion years. But Whittaker's word, protist, has come to be a synonym for protozoan, thus suggesting unicellularity despite the fact that many kinds of nonbacterial life assembled themselves into multicelled creatures without ever becoming full-fledged plants or animals. So, in our modified five-kingdom system, Whittaker's protists are known as smaller members of the huge group, the "protoctists."

23

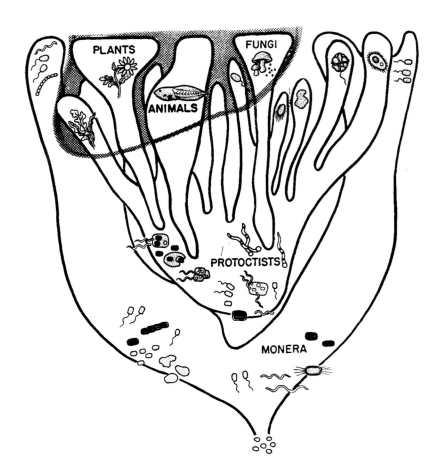

FIGURE 1.4

This phylogeny shows natural relations among organisms, members of the five kingdoms. The macrocosm is shaded. All other organisms belong to the subvisible world. (Drawing by Laszlo Meszoly.)

Inelegant, but useful, the term "Protoctista" originally was coined by British biologist John Hogg in 1860 to refer to "organisms that are clearly neither animals nor plants." It was resuscitated by Copeland and restricted to exclude not only plants and animals but also fungi and bacteria. This is the way we use the term. Protoctists include all the algae, the ciliates, and the amebalike cells in all their variation and complexity, among them the slime molds. Protists are microscopic members of the kingdom Protoctista.

The five-kingdom classification used in this book is based on the idea of Canadian microbiologist Roger Stanier that "the basic divergence in cellular structure that separates the bacteria and the blue-green algae (cyanobacteria) from all other cellular organisms probably represents the greatest single evolutionary discontinuity to be found in the present-day world." The kingdom Monera, then, actually represents a sort of superkingdom because the members of the other four kingdoms are all based on the same cellular plan. But even Whittaker's five-kingdom system might seem prejudicial because it gives eukaryotes four kingdoms, while prokaryotes are allocated only one. Because scientists belong to a eukaryotic kingdom, they cannot avoid a certain naked-eye, or macroscopic, perspective. The human bias is one of increasingly greater distinctions among organisms closest to our own size and form. Even so, the five-kingdom system is a step in the direction of clarity, a move toward equal representation of nature's diverse constituencies (Figure 1.4).

The new taxonomy helps us look at life in a less human-centered way because a given life form does not have to match our own visible world and be either plant or animal. For example, consider the relationship of sex and reproduction. Reproduction, the simple copying of cells, is a property of all known life forms. But in the bacterial kingdom, sex—the introduction of foreign genes in the form of DNA into a cell—and reproduction—the making of offspring—are entirely separate phenomena. Nearly all species of fungi and protoctists also regularly reproduce without engaging in any sexual process. In plants and animals, however, sex and reproduction are tightly interwoven processes. Animal and plant cells differentiate: the fertilized cell (zygote) produced by the fusion of egg and sperm nuclei grows to form many different kinds of cells organized into distinct tissues. Yet despite such differentiation, plants and animals must keep one immortal copy of all their genes in at least one reproducing cell. The billions of protective body cells produced each generation do the genes' bidding. Our lungs, heart, and brain all exist because historically they have protected and delivered the genes carried in egg and sperm cells to future generations.

Animals are marked by development from an embryo that forms from two parents—the sperm-forming male and the egg-forming female. Figure 1.5 illustrates the stages of development. After sperm meets egg but before the fertilized egg develops into a particular animal species, the embryo develops into a hollow ball called a blastula, seen in the center of the drawing. The blastula forms out of the cell divisions that follow fertilization of an egg carrying one set of chromosomes by a sperm carrying another set. There are about 31 major groups, or phyla, of animals. One is our own chordate phylum, which includes sea squirts, fish, birds, reptiles, and mammals. Sponges,

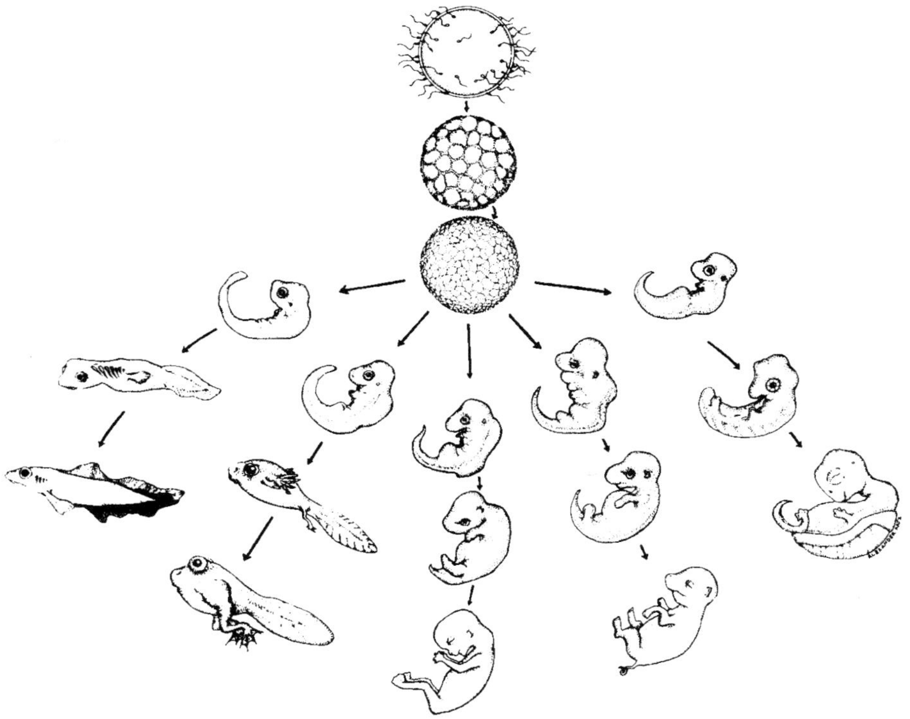

FIGURE 1.5

Animals develop from a hollow ball of cells, the blastula, which forms from the fertile egg after the union of an egg and sperm. Although many small sperm contact the large egg, they compete with each other, and only one enters. Developmental stages of members of our chordate phylum: fish, frog, human, pig, and turtle embryos. (Drawing by Sheila Manion-Artz and J. Steven Alexander.)

members of the phylum Porifera, are least like the others. They probably evolved along a path that split from the rest of the animals over half a billion years ago. Nonetheless, because they share development from a blastula that itself comes from the fusion of sex cells, sponges may be considered animals.

Plants also develop from multicellular embryos and thus cannot be one-celled. Plant embryos are not blastulas; they are a layer of "fertile" cells surrounded by supporting "sterile" tissue. However, plant embryos do form by the fertilization of plant egg nuclei by male nuclei carried by pollen or even swimming plant sperm. Some plants can self-fertilize while others must cross-pollinate, but plants, like animals, are intrinsically sexual.

Members of the kingdom Fungi include yeasts, mushrooms, puffballs, and molds. Fungi are neither plants nor animals because they grow as threads (hyphae) from spores without ever forming embryos during their lifetime. Fungi also have a unique mode of absorptive nutrition. They do not ingest the food of others like animals or make their own food from sunlight, air, and water like plants. Instead, they excrete digestive enzymes into their surroundings and absorb small molecules through the outer walls of their nucleated cells. Most adult fungi, unlike plants and animals, have cells that contain only a single set of chromosomes. Fungi also lack cilia, sperm tails, or other specialized undulating motility organelles. Such cell tails appear at some time during the life cycle of nearly all animals and plants with the notable exception of flowering plants.

Cilia, sperm tails, and other undulating cell appendages of eukaryotes are generically known as undulipodia. Flagella, which are similarly long and thin but otherwise completely different from undulipodia in their structure, are found only in bacteria. In this book, we distinguish flagella from the undulipodia of nucleated cells although, in many texts, no such distinction is made.

All organisms that do not display the characteristics of the other four kingdoms are relegated to the kingdom Protoctista. They are a diverse group,

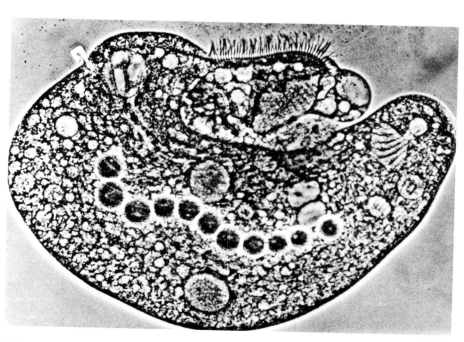

FIGURE 1.6

The protist *Stentor coeruleus* as seen under the light microscope. Such stentors, members of the kingdom Protoctista, are blue and "huge" for single cells—when they are stretched out, some are as long as 700 micrometers, nearly a millimeter long.

including all the unicellular microbes with nuclei, such as amebas, and their multicellular descendants, such as slime molds and seaweeds (Figure 1.6). The vast majority, maybe even all of them, can reproduce directly—single parents reproducing without benefit of sex. Some, at some time in their life cycle, become sexual and seek a mate; others never do.

Protoctists are still being discovered and described today. Professor John O. Corliss of the University of Maryland estimated that there are some 200,000 species of these aquatic curiosities. They are of extreme importance to both medicine and evolutionary thought. Medically, the majority of tropical diseases are associated with the presence in animal tissues of protoctists with complex and mysterious life cycles. Some must even infect several different kinds of animals in order to live. The diversity that makes protoctists dangerous in the tropics makes them interesting from the standpoint of life's history. They provide important clues to the origins of mitosis, multicellularity, and the sex lives of animals and plants.

▪ WINDOWS TO THE SUBVISIBLE ▪

Microscopes and Eyes

Before we proceed to the next chapter, a bit more needs to be said about optical instruments. Our view of the microcosm originated with the use of magnifying lenses and microscopes. Centuries ago, people must have noticed the natural magnifying properties of water droplets and calcite crystals, which distort and enlarge small objects. The Arab physicist Abu Ali al-Hazan ibn Alhasan (962–1038) was the first to record that a piece of glass, flat on one side and curved on the other, would magnify small objects. In the thirteenth century, the English philosopher Roger Bacon described how such lenses could be used as spectacles. The first dual-lensed microscope, similar to a hand-held telescope, appeared about 1590. But the potential of the compound microscope was not yet fully realized. At this time, it could only magnify from 50 to 100 times.

Optical instruments did not, and do not, always give an undistorted view of their subject. (Nineteenth-century astronomer Percival Lowell was convinced that the cracks on the surface of Mars were canals engineered by civilized Martians.) And some early critics of the microscope correctly assumed that features seen with the instrument were fabrications. The eye itself is an optical instrument, working on the lens principle. It, too, is susceptible to deception. At times eyes produce their own illusions. Some come from the blood coursing through the vessels near the retina. Close objects, such as small insects or bits of detritus on the eyelid, may be mistaken for large, distant objects. Bright, opposite colors (so-called "afterimages") appear after fields of color or lights are stared at and then removed. Perception, either through the eye or with the technological enhancement of glasses or microscopes, is always a participatory phenomenon. Comparative experience, details, and confirmation by other senses play a crucial part in what we "see." Ultimately, the shared perceptions of others and our own past perceptions provide a frame of reference that determines what we see as reality and what we dismiss, consciously or subconsciously, as artifacts or mirages.

Eyes are such finely adjusted focusing instruments that when Charles Darwin (1809–1882) carefully outlined his concept of "descent with modification" they seemed to represent an exception. Instead of having gradually evolved from simpler structures, it was easier to see them as having been created, as evidence of plan and design. Yet evidence garnered from these wondrous organs and from their extensions in the form of microscopes has revealed plausible pathways for the evolution of the eye by such "descent with modification."

All eyes have precedents in the light-sensitive membranes of microbes. The sensitivity to light already is highly developed in the tiny purple bacteria like the one at the bottom of Figure 1.7. All the little membranes inside the photosynthetic *Chromatium* bacterium are sensitive to sunlight. Another example is the light-sensitive protein, called rhodopsin, in the salt-loving *Halobacterium*. Made of a colored compound directly related to vitamin A and a large attached protein, the chemical structure of the bacterial colored compound is so like that in the eyes of animals that only sophisticated chemical techniques can distinguish between them.

Not only are light-sensitive proteins prevalent in the microcosm, but the functioning eye has repeatedly evolved from simple components, all of which are of some value to the organisms that bear them. Eyes evolved many times in different groups of organisms. A comparison of eyes in different organisms is shown in Figure 1.7. The photosensitive parts of the organisms are shown magnified toward the right side of the page. In the protist *Erythrodinium,* a sea whirler, a great part of the body is modified to see. The insect eye, the sea slug eye, and the mammal eye all have functional parts in common: light-sensitive membranes, lenses, and movable parts used for focusing.

We can trace a continuum of visual acuity from the simplest light-sensitive bacteria to today's most complex microscopes with resolutions limited only by the wavelength of the electron particle beam. Microbes can be studied live and can be seen going about their daily activities by using the light microscope. Sometimes microbes are so small and watery that their details escape observation by the light microscope. As an aid to recover the lost details, special dyes or stains are used that react in specific ways to distinguish various structures.

Gram stain, developed in 1884 by the Danish physician Hans Christian Joachim Gram (1853–1938), is used to distinguish between two major kinds of bacteria. Gram positive bacteria react positively and take up the violet color of the stain chemical, known as crystal violet. All other bacteria fail to retain the purple color and stain pink by default. Bacteria that do not hold the purple stain are considered Gram negative. Electron microscopic studies have revealed that Gram negative bacteria have a second external membrane surrounding the ordinary cell membrane of Gram positive bacteria. Whether this second membrane prevents crystal violet from staining or just keeps it from penetrating the cell wall layer that actually takes on the purple color is not known. In any case, the status of having two membranes and being Gram negative can now be determined without the crystal violet stain if, that is, you happen to have an electron microscope.

The scanning electron microscope (SEM), one type of electron microscope, works by shining on an object electron beams that have been "focused" by magnetic "lenses" in a vacuum. As the power of a microscope increases,

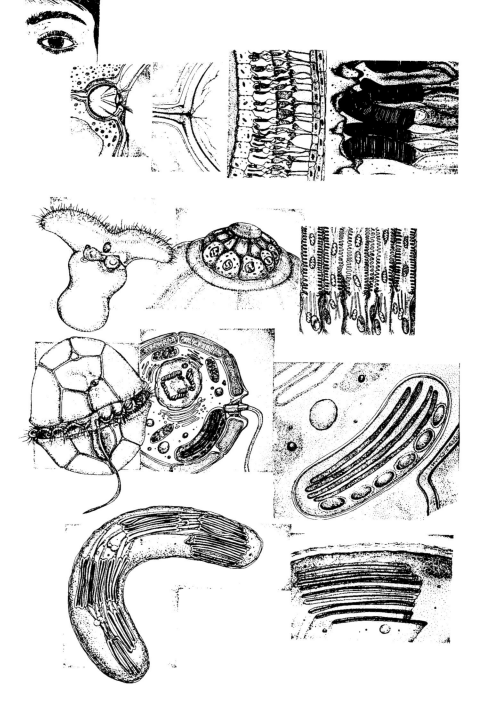

FIGURE 1.7
Light sensitivity in organisms: comparison of photoreceptors and eyes. From top to bottom: (1) human, (2) snail-like mollusk (nudibranch larvae), (3) protist (a dinomastigote), and (4) bacterium (*Chrornatium*). All are sensitive to light. The photosensitive membranes are shown at high magnifications to the right of the drawings of each example. (Drawing by J. Steven Alexander.)

however, depth of focus is sacrificed. Thus, the highest power microscopes can examine only an exceedingly thin layer of material at a time. So, in a way, Leeuwenhoek with his homemade microscope experienced the subvisible world of incessantly active life forms more fully than many modern researchers in ever more specialized disciplines, using ever more precise instruments to explore ever more exquisite details—at the price of a complete view of the microscopic universe they are examining.

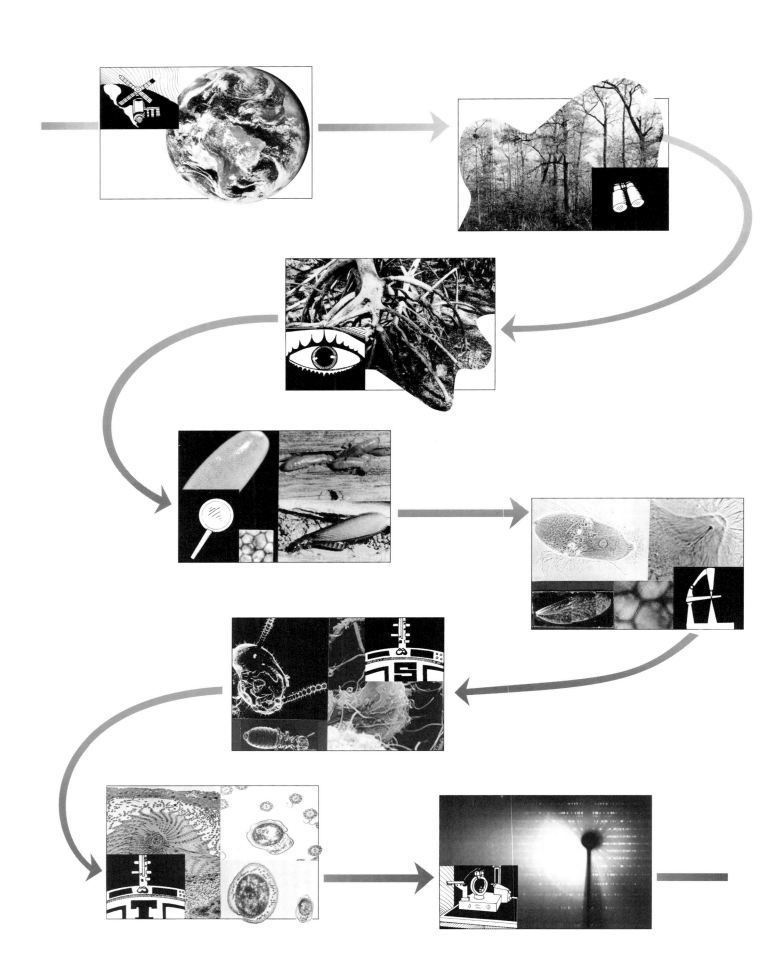

T·W·O

OUR
ULTIMATE
ANCESTRY

■ JOURNEY TO THE CENTER OF THE CELL ■

From Earth Image to a Microbe's DNA

ith all our peeking and spying into the world of microbes, we are still woefully ignorant of the nature, power, diversity, and oddities of the microcosm. Even our closest approaches to the microbial world so far have been like witnessing cities from space. The remote sensing of images may show broad changes. The contraction of green and expansion of brown on a satellite map may, for example, indicate jungles becoming desert, but nothing short of sharp focus at very low altitudes will convince us of the actual deforestation process. And even then, close-up studies of one feature may defocus others, obscuring the vast, interconnected nature of the phenomenon, be it life on planet Earth viewed from a telescope in space, or life in a drop of pond water viewed through a microscope in the biology laboratory.

The world is structured at all levels from the galactic to the subatomic. This guide to the microcosm is most concerned with the levels of biological organization from one micrometer (one millionth of a meter) to one millimeter (one thousandth of a meter) in size. There are many ways to descend into the subvisible universe we call the microcosm. The following is one particular journey, a voyage into the center of a living cell, that starts from outer space and ends inside a termite biochemical (Figure 2.0). We have chosen to voyage into the termite hindgut world because it is a well-bounded microworld about which something is known. All organisms are part of a symbiotic network that spans from the sea floor to the atmosphere, including the cells of our bodies and encompassing the entirety of life. The existence of this network becomes clear as we zoom in from our initial point in space to concentrate on a single cell within a southern Florida termite, increasing the powers of magnification as we descend to view the "heart" of the microcosm.

FIGURE 2.0
Visual journey from space to the center of a cell. Descent into the microcosm: Satellite camera, binoculars, eye, magnifying glass, light microscope, scanning electron microscope, transmission electron microscope, and X-ray diffractometer. Each visual instrument corresponds with an image: the Earth, a Florida mangrove forest, mangrove roots with termites, a colony of termites (*Kalotermes schwartzi*), a termite, a termite egg, microbes from the hindgut of one of these termites, a "large" microbe from a termite called *Trichonympha ampla*, undulipodia, and an X-ray photograph revealing molecular structure of an organic compound. The letters refer to the sizes of objects generally seen by each instrument (in meters).

31

From a satellite's camera (Figure 2.0), we look at Earth through an atmosphere surprisingly rich in such gases as oxygen, methane, and ammonia. It is surprising because these gases are highly reactive with each other; they persist in their relatively high ratios on Earth through the vehicle of life. These gases are virtually absent on our lifeless planetary neighbors Mars and Venus, which are enveloped in atmospheres mainly composed of unreactive carbon dioxide. Our atmosphere is not a sterile backdrop to the evolution of life. It is an extension of the biota, of all living organisms at the surface of the globe. Like the shell of an oyster or a bird's nest, our environment is largely produced by the combined activities of all life.

As we near Earth's surface, entering the optical range of a pair of binoculars (Figure 2.0), the view becomes more familiar. We can see the tangled floor of a mangrove forest on the coast of southwestern Florida. Here are some of the trees that emit the oxygen we detected from our vantage point in space. The forest is rich in associations, pacts, and permanently merged organisms. It is steeped in rivalries, predations, deceptions, even miniature arms races as prey organisms evolve new biological chemical weapons while their predators counter with new means of detoxification. Sometimes organisms become so dependent upon each other it is difficult to tell if they are competing or cooperating. Symbioses—the intimate sharing by members of distinct species of everything from foods to physical bodies and even metabolic pathways—is a hallmark of the Earth's biota. No square centimeter of soil is free from bustling communities of billions of microbes.

As our scale shifts once again to that of ordinary eyesight (Figure 2.0) we look down at the ground. Salt marsh plants and the aggressive aerial roots of red mangrove trees are plainly visible. The tiny dots on the shaded roots are termites scuttling in and out of the moist wood. At the level of a magnifying glass (Figure 2.0), the different castes of termites come into view. We see the larger winged forms (the bridal alates), ready to embark on their nuptial flight to begin new colonies, and the smaller, far more numerous worker termites, chewing away at the wood that serves both as a nest and as food. From a closer range we can even see an oval termite egg about one millimeter long. Focusing on the egg's surface with a microscope reveals the hexagonal latticework of its coat.

Our journey continues inward with the revelations of the scanning electron microscope (SEM). The SEM is special in its ability to scan images both at the level of the ordinary light microscope and at far greater magnifications. Able to browse through the layers of cells of the termite body, the SEM is an excellent tool for the professional biologist. The instrument shoots a beam of electrons rather than the visible light of photons, revealing the intricate detail of the three- or four-millimeter-long termite. Then, as the magnifying power increases, the SEM treats us to a view of the surface of the inside of the termite's hindgut, a swollen part of its intestine (Figure 2.0). Here, sealed off from the oxygen of our world, live dense and complex communities of interacting microorganisms. The hindgut microbes are not simply pesky guests, but an integral part of the termite. They are indispensable to their host's digestion of cellulose, the tough, starchlike polymer of wood (Figure 2.1).

32

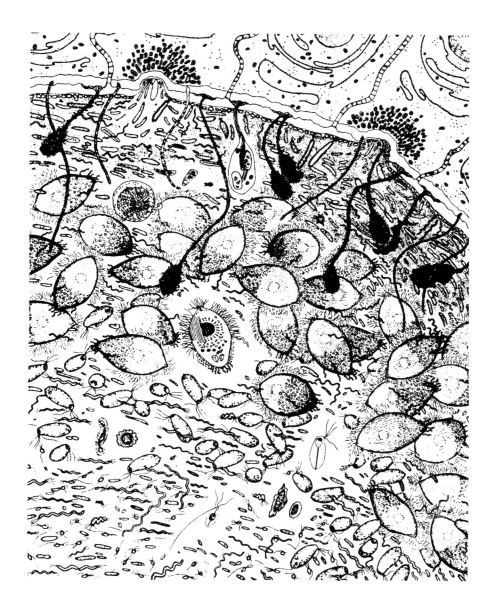

FIGURE 2.1
The microbial community of the termite hindgut. Christie Lyons, the artist, has reconstructed what the termite microbes look like alive and living together. Some (those with halos of light around them) show detailed structure seen only with transmission electron microscopy. Parts of five termite gut cells and their septate (striped) connections can be seen at the top.

The microbial community of the termite hindgut is inherited not through the egg but by force feeding from nest mates in each generation. The rear end of a worker termite is presented to the mouth of a newly molted or newly hatched termite. The microbes, ingested by newly hatched termites, are quickly passed along to nest mates in this process, known as proctodeal feeding. The termite soldiers, too well armed with their prickly jaws to eat wood in the normal fashion, depend on such backward "feedings." Wood particles and microbes carried in the hindgut fluid are routinely distributed to soldiers in this way.

The scanning electron microscope next focuses on the wriggling bacteria—spirochetes and so-called rods—that live in the oxygen-poor environment of the hindgut. The spirochetes resemble animated pieces of fusilli pasta. Their anaerobic environment may in fact re-create the oxygenless condition of early Earth. By revealing the world of the very small, the SEM

33

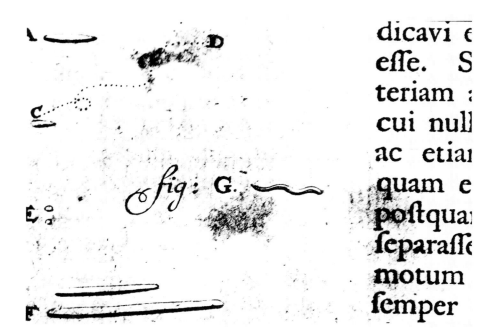

dicavi ε
eſſe. S
teriam ι
cui null
ac etiaι
quam e
poſtquaι
ſeparaſſε
motum
ſemper

has introduced us to a type of life form that dominated the earth for 80% of geological time prior to the appearance of cells with nuclei.

Now we jump back to the realm of the ordinary light microscope (Figure 2.0). Here we are capable of seeing through the bodies of live, watery microbes and into the cells of animals and plants. It was with this instrument that Anthony van Leeuwenhoek discovered the existence of bacteria. Leeuwenhoek himself may have been the first person to see and report on spirochetes (Figure 2.2) when he wrote:

> *"I have also seen a sort of animalcule that had the figure of our river eels: these were in very great plenty and so small withal that I deemed 500 or 600 of 'em laid out end to end would not reach to the length of a full-grown eel such as there are in vinegar. These had a very nimble motion, and bent their bodies serpent-wise and shot through the stuff as quick as a pike does through water."*

Also found in the hindgut ecosystem is the protist *Trichonympha*. *Trichonympha* is a creature that excites the mythological imagination for it has a split personality: it is not one, but several different kinds of microbes living together. Some bacteria live inside its nucleus, while other kinds squirm at its surface. Corkscrew-shaped spirochetes frequently are found attached to the rear surface of the *Trichonympha* collective. Their wavelike motion helps bits of wood enter the protist's hind end. Apparently *Trichonympha* feeds on wood particles while its adhering spirochetes gorge on something else, probably some form of *Trichonympha* waste (Figure 2.3).

Now we look even more closely at *Trichonympha* using the transmission electron microscope (TEM) (Figure 2.0). This tool, which also relies on streams of electrons, is even more powerful than the scanning electron microscope. The insect hindgut, cut very thin, is embedded in plastic. When slices are placed in the TEM, then we see the fine structure of the

FIGURE 2.2

Did Anthony van Leeuwenhoek see a spirochete? The illustration on the right from Leeuwenhoek's *Arcana Natura Detecta* was photographed by R. Guerrero, the photo above is a negative-stained electron micrograph of a spirochete.

Trichonympha cell itself, including the central nucleus. Rows of cell structures, organelles called undulipodia, are clearly visible on the surface. In a cross-section of *Trichonympha,* these "tails" are seen as wheellike structures. Each undulipodium contains nine pairs of slender, round filaments called microtubules (they look like tiny coupled circles in the cross-section), surrounding a central pair. The whole arrangement looks like the dial of a rotary telephone. On the right in Figure 2.0 we see slices through two sorts of similar-looking circles. The larger ones are the helical spirochetes mixed in with *Trichonympha*'s own undulipodia, helping to propel the protist along. The photo on the left shows the smaller microtubule circles of the actual undulipodia.

From an evolutionary standpoint, undulipodia may have once been independent undulating spirochetes. This idea implies that the tail of a sperm cell —an undulipodium—evolved from a wriggling spirochete. At present, the idea is only an unproven hypothesis. Nonetheless, the hypothesis finds support in the fact that antibodies reactive to the undulipodia of eukaryotes (as in sperm tails) also react specifically to certain spirochete bacteria. This suggests there is more combining the two types of structures than meets the eye. We summarize the spirochete hypothesis below as a possible explanation for the development of undulipodia. The attractive part of this hypothesis is that it helps make sense of the great evolutionary gulf separating prokaryotes, non-nucleated cells that never have undulipodia, from eukaryotes, nucleated cells that generally possess them.

Spirochetes may have entered associations by forming attachment sites at the edge of other bacterial cells. Perhaps originally pathogenic, like the spirochetes that cause syphilis, some of them may have moved from the surface of larger cells to the inside. Eventually, their locomotory mechanisms could have become the microtubule systems of mitosis and meiosis. Scien-

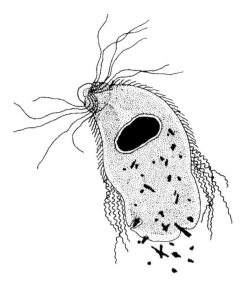

FIGURE 2.3
Trichonympha, a wood-eating protist found in the hindguts of termites. The black particles are tiny wood chips. (Drawing by K. Delisle.)

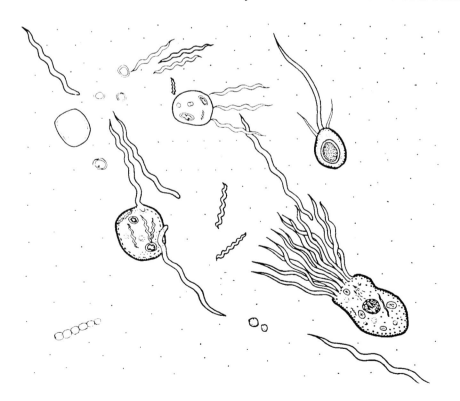

FIGURE 2.4
Spirochete motility symbiosis. Spirochetes attach to and help move other organisms.

35

tists are still mystified by the widespread appearance of these "telephone dial" structures in so many electron micrographs. It makes sense that they would have a common ancestry. Perhaps sperm tails, cilia, and other organelles that exhibit this [9(2)+2] microtubule ultrastructure (making them, by definition, different forms of undulipodia) originated as spirochetes. The spirochetes, with their own microtubules (some, perhaps, in a [9(2)+2] array), could have become so completely dependent on and valuable to certain protists that they merged with and changed their hosts completely. Their wriggling and squirming could have evolved into the constant streaming and complex transportation systems found inside all eukaryotic cells today (Figure 2.4).

The final optical level of our journey is so detailed that it is more mathematical than visual. At the bottom right of Figure 2.0, an organic molecule from a cell has been crystalized and subjected to X-rays in an effort to detect its structure. The instrument, known as an X-ray diffractometer, records the angles and quantity of diffracted X-ray beams. By bouncing X-ray light with its very short wavelength off molecules, their structure can be determined. The result, interpreted by a computer, is shown in the final picture. The carbon and other atoms that make up DNA were precisely located with this same technique in the famous discovery of the double helical structure of that molecule.

As we can tell from our journey into the microcosm, intricate community structure is a pervasive phenomenon from the largest to the smallest levels. The whole Earth and the tiny protist cell both are products of living together. The *Trichonympha* need their associated bacteria. We need our intestinal bacteria to help digest our food and provide us with vitamins. We would die without them unless supported by intensive care. Microbes also digest the mucus produced in the membranes of our intestinal walls, which our body's enzymes cannot break down. Without microbes, large amounts of mucus accumulate in the intestine, the mucus attracts water, and the intestine becomes dangerously swollen. Beside such helpful microbes, every one of our uncounted billions of body cells (estimates range as high as 10 quadrillion), descending in each generation from a single cell, has a multiple ancestry. Each body cell that derives from the union of sperm and egg has ultimately evolved from the crowded cooperation of different kinds of bacteria. Although this sounds more like science fiction than sober scientific reality, the idea that the eukaryotic cell represents some sort of bacterial merger is now accepted as a fact of the history of life on Earth.

As life continues, there is little doubt that novel organisms will evolve and recombine. Associations between organisms—including those endosymbiotic mergers in which one form lives inside the borders of another—probably will continue to play an important role in the genesis of new life forms. In the past era of microbial evolution, symbioses have been as crucial as the competitive struggle to survive. Observing today's bacteria, protoctists, and fungi with optical instruments in the laboratory and in the field helps us to imagine the time when microbes dominated the Earth. In the following sections, we will take a rapid tour of the history of subvisible organisms, a history which set the stage for the emergence of the generally larger plants and animals. Not only is the history of the microcosm fascinating in its own right but, to this day, microbes underlie all the crucial ecological cycles of our shared planet.

A Subvisible Heritage

Our voyage from the superorganism of the biota to molecules of organisms living within other organisms suggested the degree to which more or less intimate associations among living organisms are central to life on Earth today. As we trace back the spiral of evolution (Figure 2.5), we find that symbiosis played an even more prominent role in forming the communities of bacteria that became the ancestors of our cells.

Most lines of speculation consider life to have been the result of chemical associations. Autocatalytic precursors to life—in other words, self-assembling and self-reproducing chemicals that were not yet embodied in living beings—may have joined to produce the first life forms. Preliving (prebiotic) chemical reactions, which included the cyclical self-synthesis of the macromolecule RNA, probably preceded the first true bacterial cells. Today, RNA's major job inside all life forms is to turn the message contained in the

FIGURE 2.5

Spiral of evolution from the origin of the universe. Among those shown are a lemur, mating hypermastigotes, *Latimeria*, a coelocanth, *Stegosaurus*, origin of the solar system, mating ciliates, blastula, DNA, *Brontosaurus*, *Homo erectus*, spore, stromatolite, seed-fern, *Ankylosaurus*, club fungi, Earth from space, oxygen bubbles, brown algae, lunar inhabitant, ecosphere, trilobite, euglenid, a polychete worm, and a future platypus swamp crocodile. (Drawing by J. Steven Alexander and Sheila Manion-Artz.)

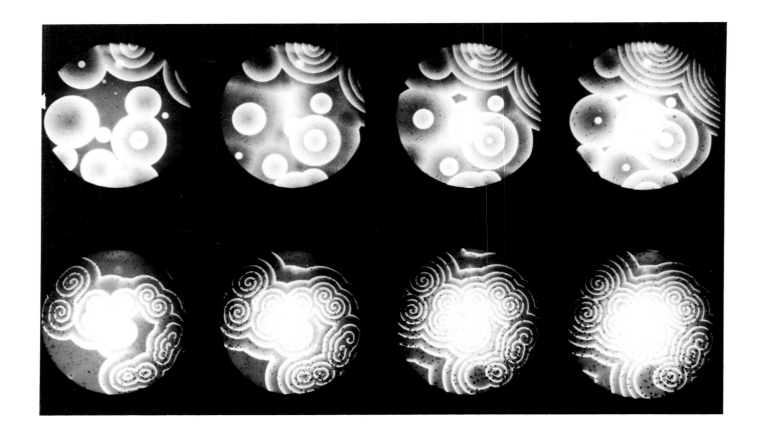

FIGURE 2.6

Increasing complexity in an autocatalytic reaction, known as the Belousov-Zhabotinsky reaction. These patterns develop when malonic acid is oxidized by bromate in the presence of cesium ions in a shallow dish. Self-organization, clearly a property of life, is also seen to be a property of certain non-living chemical systems. (Photo from A. Winfree in "The Self-Organizing Universe," by E. Jansch.)

chemical structure of DNA into the proteins that make up a living organism. At the same time, RNA makes more DNA. The first bacteria may have resulted when chemical systems composed only of RNA and protein, isolated from their environment in droplets of fatty substances, began to interact. An autocatalytic chemical reaction—one that grows progressively more complex despite the absence of any life form (or, in this case, of any biochemicals at all)—is illustrated in Figure 2.6.

Of course, lifelike and living are two different things. Theoretically, it is impossible to prove how life began unless future science evolves to a point where it can predict every future action and thus retrace every past one. Perversely, the origins of life and matter seem knowable in inverse proportion to our curiosity. But let us consider them the original assumptions, the givens. Like the origin of the universe, the origin of life also may be intrinsically unknowable. Even re-creating it in a test tube would not prove that it began that way in the first place.

Nonetheless, we can recount a story of our past. Although it sounds strange, it is believable and, unlike the great majority of myths, it is completely consonant with science. This does not mean it is the absolute truth, however. The story will certainly change to keep up with incoming details; it will be updated in order to conform with new scientific results. It is good to bear this in mind as we set the four-billion-year-old stage for our description of modern bacteria, protoctists, and fungi.

Let's begin with the beginning of life. We may reasonably assert that members of what might be called the macrocosm—that is to say, plants, animals, large fungi, and other obvious, recent creatures—would never have

38

survived in the early world of microbes. First of all, for over a billion years from the origin of the first cell there was no oxygen available in the breathable form (O_2) in the atmosphere. (Of course, oxygen as the "O" in the H_2O of water was indeed present. But this was totally useless for supplying oxygen needed for breathing.) Oxygen as the reactive gas O_2 was bound up in such compounds as carbon dioxide and water vapor, or steam. Another not-so-minor difficulty for animals and plants would have been the hostile landscape of the early Earth, which was bombarded by planetesimals and scorched by liquid rock erupting from volcanos.

Earth probably formed from bits of matter coalescing under the powerful influence of gravity. Heavy elements, such as iron and nickel, made their way to the center of the proto-earth, while lighter ones tended to float at the surface. The gases of the original solar nebula contained not only such elements as iron, which quickly became solid as temperatures fell, but also substances that remained gaseous at lower temperatures. These latter elements—hydrogen and noble gases such as argon, neon, and xenon—along with carbon monoxide, carbon dioxide, and nitrogen formed the earth's first atmosphere. This atmosphere was largely made of what was left over from the condensation of the solar nebula into planets, moons, and the Sun. Like the Sun, it was rich in the lightest and most abundant compound in the universe, hydrogen.

Could life have persisted through the hellish primeval storms? Even the hardiest microorganisms might succumb under such pressures. Yet some, such as *Bacillus thermophilus* ("heat-loving bacterium"), which reproduce in the scalding temperatures of Yellowstone National Park hot springs, raise intriguing doubts. Indeed, virtually as soon as there could be any evidence of life from the rock record, there is. In three-and-a-half-billion-year-old sediments from the Barberton mountains in South Africa and the Warrawoona rock formation in North Pole, Australia, putative bacteria left marks in the shape of tiny spheres and filaments. Rounded rocklike domes called

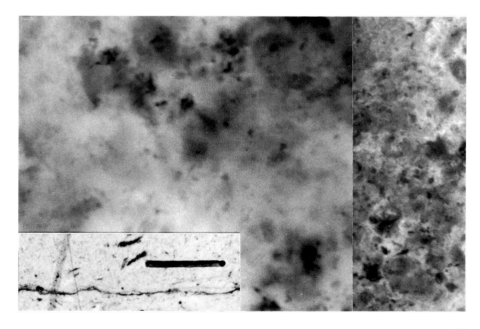

FIGURE 2.7
Microfossils in a thin-section of chert, taken from the Swartkoppie zone of the Swaziland formation in the Barbedon Mountains (town of Fig Tree), S. Africa.

stromatolites grew in western Australia. Stromatolites are made when generations of photosynthetic bacteria live on top of each other, divide, and squirm upwards toward the light to make new layers of their crowded bodies. Both ancient and modern stromatolites are produced in the same manner, by lush, brightly colored, tightly packed layers of aggregated bacteria.

Taken from southern Africa, a smooth form of three-billion-year-old black quartz, called chert, that contained a small amount of carbon was cut into microscopically thin sections and examined under the light microscope. The result, as you can see in Figure 2.7, gives evidence of early reproducing cells—some, it appears, captured by hardening muds in the act of division. If we go back further in the rock record and look at the very oldest rocks in the world, which come from Greenland, we find no convincing evidence for life. But at this time, no good fossils could have been left because the Earth's crust, partially molten and indented by meteorite impacts, was still being exposed to the severe heat and high pressures that would probably destroy all traces of life.

■ THE ENERGY REVOLUTION ■
Photosynthesis

By all accounts, the first life forms must have been bacterial. Even though bacteria are among the smallest organisms known, their direct remains can be found as fossils in the geological record. Bacteria microfossils of the same size and shape as modern forms, both those found in chert rocks from South Africa and others from western Australia, have been dated older than 3000 million years. How did these microbes survive the steamy, scorching oxygenless past?

Like modern anaerobes that proliferate in muds and inside other organisms—anywhere free of oxygen—the earliest bacteria did not need oxygen. Indeed, to them it was a poison. The first bacteria are thought to have been fermenters that took their nourishment directly from their surroundings. They derived energy by converting food—organic compounds such as sugars—into other organic compounds such as alcohol, just as winemaking microbes do today in the airless environment of champagne bottles. But where did the first fermenters, lacking freshly squeezed grapes on which to feast, obtain their nutrients? They probably used not the biochemicals of life forms, but carbon-rich compounds floating in water that were formed directly in the environment by the Sun's ultraviolet radiation. Such radiation tended to break apart chemical bonds, causing elements to regroup into prebiotic "candy."

But, since they were gobbled up faster than they could be replenished, the sun-baked treats did not last. The alternative to finding a new source of food was death by starvation. The greatest step in the evolution of life was taken when the first photosynthetic bacteria evolved. Photosynthesis is life's ability to absorb visible sunlight and use the solar energy to make food directly from the carbon dioxide of the atmosphere. No skyscraper, no computer, no brain rivals the accomplishment of photosynthesis—the underlying layer of support upon which the rest of the biosphere's energy needs are based.

Besides carbon, hydrogen is essential for photosynthesis as a source of electrons to power the process. The first microbes that grew, spreading across the warm shallow seas and often molten lands of early Earth, probably were similar to the green and purple photosynthetic bacteria that thrive today in oxygen-poor environments. At first, the early photosynthesizers probably took their hydrogen straight, as a simple gas emitted directly through the Earth's crust into the atmosphere. But soon after, the Sun itself made this impossible.

The Sun, according to most theories of stellar evolution, was once considerably cooler than it is today. As it pulsed radiation through the solar system, turning hydrogen into helium in its vast nuclear reactor of a core, the Sun burnt off the light gases of Earth's original atmosphere, sending them into space. The planet's feeble gravity failed to retain the hydrogen, and its supplies diminished. Early photosynthetic bacteria were robbed of their storehouse. Meanwhile, continuing volcanic activity brought up a secondary atmosphere, consisting largely of water, nitrogen, and carbon dioxide, from the Earth's interior. As on Mars and Venus, carbon dioxide (CO_2) belched from the interior helped make a new atmosphere on Earth. Even today, the atmospheres of these neighboring planets still are more than 90% CO_2. Why is it that of these three planets only our atmosphere contains so little carbon dioxide? The answer is part of the story of life.

Hydrogen sulfide is another gas belched up through the mouths of volcanos. Accumulating as the atmosphere, and life itself, evolved, hydrogen sulfide was cherished by the oxygen-hating but sun-loving green and purple sulfur bacteria as an alternative source of hydrogen. Today, as in the remote past, in sea muds and lakes, layers of green and purple sulfur bacteria thrive on whatever hydrogen sulfide they can scavenge. But, in early times, they must have consumed up the hydrogen as it appeared so that, far from being solved, the hydrogen crisis was exacerbated. Plundered by burgeoning populations of photosynthetic bacteria, sources of hydrogen sulfide became vanishingly rare.

The squandering of the hydrogen sulfide reserves may have precipitated the next great event in the history of our planet. Photosynthetic microbes, evolving on an oxygen-poor globe, discovered water. It had been there all the time. Indeed, bacterial life had been immersed in it from the beginning. But only now, perhaps 100 million years before the time the oldest bacteria died and were preserved as fossils, did microbes begin to use water, H_2O, in a unique way: as a source of hydrogen.

The discovery of water meant that suddenly the raw materials already used as a bathing solution by all life could now be used for photosynthesis and cell growth. The oceans with their water, and the sky with its carbon dioxide, became the limits. Life proliferated but, in cleaving the hydrogen atoms it needed from the oxygen atoms of water, life released oxygen in the now breathable form of O_2 as a sort of biochemical exhaust. In doing so life created a problem for itself of daunting proportions.

Oxygen, an abundant element in the cosmos, had certainly been present on Earth since the origin of the planet. But oxygen gas reacts immediately, even explosively, with such elements as hydrogen, carbon, and iron, to form oxides such as water, carbon monoxide and dioxide, and iron ore (hematite and magnetite). Oxygen never accumulated in the atmosphere until the great proliferation of cyanobacteria produced surplus oxygen beyond that used in the reactions making water, carbonate, iron ore, and other oxidized chemicals.

Prior to the global blue-green bacteria takeover about 2.5 billion years ago, oxygen waste seeped into the air and reacted with many elements to make various types of oxygen-rich compounds. Elements such as iron, uranium, and sulfur at the Earth's surface show little oxidation before two billion years ago. Observations by geologists suggest that the Earth's atmosphere was relatively free of oxygen until this time. In the current, most straightforward scenario, it was the worldwide proliferation of blue-green bacteria that dramatically oxidized the Earth's crust and atmosphere, changing the planet forever. About this time geological "red beds"—huge rusty piles of uniformly oxidized iron minerals—began to form all over the planet. The red-bed rocks accumulated then, as they do now, in an environment abundant in oxygen gas—the same oxygen that originally was a blue-green bacteria waste.

Another indication of early life are the banded iron formations, which contain all the world's commercial iron ore for making cast iron, steel, and other alloys (Figure 2.8). The deposits formed on Earth in great abundance prior to the appearance of red beds. Consisting of distinctive layers of oxidized (hematite) and less-oxidized (magnetite) forms of iron, the presence of banded iron formations on Earth may be as much a sign of early life as the presence of coral reefs is of later, animal life. The bands seem to suggest fluctuating cycles of oxygen-rich with oxygen-poor environments. With blue-

FIGURE 2.8

Sample of banded iron formation (BIF) about 2 billion years old from Negaunee Iron Formation, Marquette, Michigan. The dark bands are hematite while the light bands are jaspilite, a chert which contains reddish hematite particles. This piece of rock is so intensely metamorphosed (from magnetite and hematite to jaspilite) that even if there were fossils in it they could not have survived the high temperatures and pressures. (Photo courtesy of H.L. James, U.S. Geological Survey Professional Paper 570, p. 49, 1968.)

green bacteria in the picture, such cyclical environments are easily explained. The blue-green bacteria could have released oxygen on a daily or seasonal basis over many millions of years. In times of relative cold and darkness—winter, for example—less oxygen would have been produced by bacterial photosynthesis. The active growth of blue-greens would have created the more oxidized (hematite) bands.

If banded iron is *ipso facto* evidence for life, perhaps the date of the oldest life must be pushed back still further. The oldest banded rocks on the earth are found in southwestern Greenland and northeastern Labrador. These distorted, banded iron formations were subject to such high temperatures and pressures that any fossils in them would have been destroyed. Nevertheless, the banded rocks require regular, repeated exposure to concentrated levels of oxygen, and oxygen was the waste of the primeval cyanobacteria. So these crushed bands indicate that life on earth, as bacteria, may even be as ancient as 3900 million years.

The accumulation of oxygen had another major effect on the Earth and, some think, on life itself: the buildup of an ozone layer in the upper atmosphere. Ozone, formed from three oxygen atoms, is important because it deflects much of the Sun's ultraviolet (uv) radiation back into space. It is a common scientific view that the appearance of atmospheric oxygen prompted animals to evolve and life to come onto land. Some say that ozone would have provided a shielding "umbrella" against uv radiation. To back this assertion, it is pointed out that the accumulation of oxygen made a planetary ozone shield only about 600 million years ago—the time of the appearance of the first good fossils of animals with shells.

But the idea that the paucity of oxygen kept life off the land seems doubtful. First of all, although the argument has been advanced that plants (and, therefore, animals feeding on them) would not have endured an ultraviolet-irradiated environment, botanists have pointed out that the major obstacle in the evolution of plants was desiccation, the threat of drying up and blowing away. Without waterproofing measures, the level of uv radiation wouldn't have mattered much anyway; plants would have dried to brittle flakes even under ordinary light.

A second argument against oxygen being crucial to the development of life on land comes from geology. Dating the rust deposits known as red beds indicates that atmospheric oxygen, the rusting agent, had already accumulated in substantial amounts by some 1.8 billion years ago. Thus, an ozone "shield" would have been in place well before the evolution of animals some 700 million years ago, and uv radiation wouldn't even have been a factor in deterring the spread of organisms to land.

Furthermore, experiments subjecting both anaerobic and oxygen-loving bacteria to nearly lethal doses of ultraviolet light reveal many mechanisms of protection, including moving into the shade, self-repair of genetic damage, and hiding behind uv-absorbing salts and other chemicals. Even ultraviolet radiation corresponding to that of the early Sun, pulsing through an atmosphere in the total absence of ozone, did not kill resistant bacteria living in natural environments that provided shade and uv-absorbing chemicals. Anaerobic *Clostridium*, for example, is very resistant to ultraviolet attack. Oxygen-loving *E. coli*—capable of growth by fermentation in the absence of O_2—survives better when growing under doses of ultraviolet in the absence of O_2 than in the presence of that gas. (Uv light is reactive with cell material when there isn't any oxygen around; it is doubly dangerous in the presence of O_2. Both situations can destroy cells by disrupting chemical bonds and thereby metabolic pathways necessary for life.)

Anaerobic bacteria have kept mechanisms of ultraviolet protection viable for billions of years. In fact, a major effect of exposing bacteria to ultraviolet radiation is an increase in their likelihood of engaging in genetic transfer. This may be useful because it produces new strains of microbes, which may combine important traits to survive suddenly changing, dangerous environments. Uv radiation also "turns on" bacterial DNA repair systems. Bacteria have always had ways of dealing with ultraviolet light, such as seeking shade, donning "hats" and "sunglasses," or building skeletons.

One more point on ozone: the primary danger of the depletion of the ozone layer, of which much has been made in the popular press, is the contraction of skin cancer and weakening of the immune system in light-skinned beach-goers. Far more serious may be other human activities such as destruction of the rain forests (which contain the largest concentration of species diversity on the planet) building on marshlands (which may upset the global atmospheric balance), and polluting the air, oceans, and land with toxins other than the chlorofluorocarbons (that may affect the ozone layer).

▪ THE MERGING BACTERIA ▪

But the enigma of the transition from bacteria to animals is connected to the appearance of atmospheric oxygen. Most multicellular organisms, including multicellular bacteria and cyanobacteria, as well as protoctists, such as the plasmodial slime molds, eagerly breathe oxygen. Yet oxygen alone, although necessary to the appearance of animals and plants of the microcosm, was not enough. The evolution of animals from bacteria depended first on bacterial community living, on microbial partnerships and symbioses. Without symbioses there would have been no early animals at all.

The word *symbiosis* derives from the Greek words meaning "living together." Everyone agrees that people must learn to live together in new ways if we are to survive on this increasingly crowded planet. But all people belong to the same species, *Homo sapiens*. Symbiosis refers to the living together of individuals who are members of different species, a close relationship between or among different kinds of organisms through significant portions of their life cycles. The ability of people to coexist with millions of other species of life on the planet today presents a major challenge. A similar challenge was met a billion and a half years ago in the eventful transition from bacterial to other forms of life.

Modern biology has confirmed Edouard Chatton's original suggestion that the most significant taxonomic distinction among life forms today is not between plants and animals but between cells with nuclei—eukaryotes—and cells without them—prokaryotes, or bacteria. In 1910 the Russian K.S. Mereschkovsky expressed the idea, then outrageous to most biologists, that nucleated cells evolved from mergers, from symbioses, among bacteria. The American biologist Ivan Wallin was a lone voice in an academic wilderness when he expressed a similar sentiment in his book *Symbionticism and the*

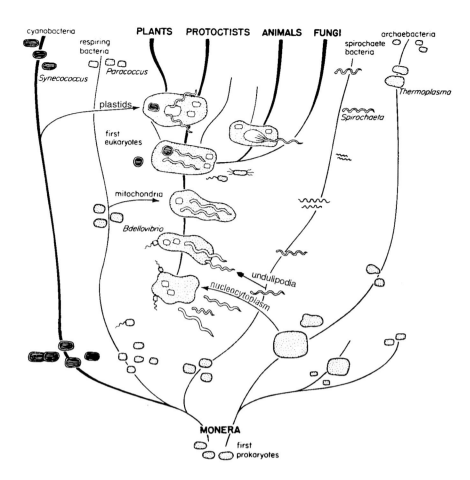

cyanobacteria
Synecococcus
respiring bacteria
Paracoccus
plastids
first eukaryotes
mitochondria
Bdellovibrio
undulipodia
nucleocytoplasm

PLANTS PROTOCTISTS ANIMALS FUNGI

archaebacteria
spirochaete bacteria
Thermoplasma
Spirochaeta

MONERA
first prokaryotes

FIGURE 2.9

Symbiosis and the evolution of the five kingdoms. This phylogeny depicts the major events in the symbiotic theory for the origin of eukaryotic cells. Respiring and spirochete bacteria were probably incorporated into an archaebacterial host, which led to protoctists, animals, and fungi. Phototrophic bacteria eaten but not digested by these became chloroplasts in algal cells. Some algal cells became the ancestors of land plants. (Drawing by Laszlo Meszoly.)

Origin of Species (1927). Today, due to the revelations of genetics and microscopy, few scientists doubt that bacterial symbiosis has been essential to the development of cells with nuclei (Figure 2.9).

The oxygen-using parts of eukaryotic cells—the mitochondria—most likely evolved from aerobic bacteria, which coped with and whose descendants eventually required the oxygen gas produced by blue-green bacteria. Every cell in the human body harbors mitochondria. Usually hundreds of these oxygen-respiring organelles can be found in a given eukaryotic cell. Although the appearance of animals was delayed until 80% of evolution had already taken place, the ancestors of your body's mitochondria probably appeared on Earth sometime before the formation of red beds, about two billion years ago.

The bacterium *Bdellovibrio* may be a modern-day relative of the type of microbe that became mitochondria. *Bdellovibrio* is a predator—a tiny, fast-moving bacterium that requires oxygen. *Bdellovibrio* bacteria invade other bacteria, multiply inside them, and finally burst out, piercing the membranes of the host cells when they have reproduced in sufficient numbers. *Bdellovibrio* bacteria then swim away and attach to new victims, entering them and beginning their life cycle again. Two billion years ago, *Bdellovibrio*-like bacteria may have entered hosts but, instead of sacrificing them, stayed in place, reaping the benefits of shelter and enabling their hosts to cope with an oxygenated environment. Even today, mitochondria possess their own DNA and replicate independently of the surrounding cell.

45

Two billion years ago, when oxygen had reacted with all the gases and minerals it could on the Earth's surface and had nowhere else to go but up, it posed a terrible threat. Oxygen was poisonous then, as it still is to all organisms today at concentrations above those to which they are adapted. The accumulation of oxygen led to the evolution of new metabolic pathways. Organisms evolved that could tolerate, and later that could use, oxygen gas. These clever microbes turned a terrible threat into a terrific advantage. By slowly burning oxygen in the cellular process known as aerobic respiration, nearly 18 times as much energy could be derived from a single sugar molecule as by fermentation.

Organisms so efficient in Earth's new oxygen atmosphere were bound to survive. But the ancestors of mitochondria did more than survive: they positively thrived. By invading but not killing larger microbes, the former predators not only provided their larger hosts with energy, but also may have protected them from the ravages of oxygen. The internal respirers in return received cellular shelter and dependable supplies of food, such as fragments of sugar molecules. This eventful partnership, a two-billion-year-old living corporation, forms the basis for all animal cells.

Cells of plants share with animals the same basic eukaryotic structure. Both have nuclei and resident mitochondria, yet plants have an added advantage: their cells are filled with grass-green organelles called chloroplasts. Chloroplasts are the green dots, the photosynthetic components of plant cells. Much evidence has accumulated in the last 20 years proving as conclusively as possible that chloroplasts are descendants of greenish, photosynthetic bacteria. For example, the green bacterium *Prochloron,* a spherical oxygen-producer that can be scraped from the surface of sea animals known as didemnids, has a lifestyle and physical characteristics that exactly match what would be expected from a free-living chloroplast if it existed.

"Plastid" is the generic term referring to the photosynthetic reddish organelles of crypter algae and red seaweeds as well as to the green photosynthetic parts of plants. All algae plastids also are thought to have come from photosynthetic bacteria. And indeed, the plastids of certain protoctists are more similar genetically to certain bacteria than they are to the rest of the algae and seaweed cells in which they reside. Studies of the component molecules making up the very long biochemicals DNA and RNA indicate that some photosynthetic bacteria in nature are extremely similar to plastids. One good interpretation is that some photosynthetic bacteria reproduced slowly in freedom while genetically closely related kinds were engulfed by larger cells, reproducing far faster in a state of permanent protective custody.

It now seems that bacterial symbiosis in the form of eukaryotic cells set the stage for the appearance of plants and animals, of larger, younger forms of life. These "macroorganisms" are recombinant mergers of members of the microcosm. They, and we, have grown from the microcosm: we contain microbes, both in their genetically altered state as organelles, and in their original state, as symbionts of our stomachs and on the surface of our large intestines, skin, and so on. As the various bacteria came together in the form of eukaryotic cells, a new range of possibilities was opened. Communities of interacting bacteria led to the visible world of plants and animals.

In retrospect, we see that the tiny members of the microcosm did more than alter their own environments. They created their environment. Their rapid growing and dying and reworking of the environment made their surroundings habitable. Ultimately, they altered the whole surface of the planet, oxidizing it. The carbon dioxide, which on Venus forms 98% and on Mars forms 95% of the atmosphere, makes only 0.03% of ours. Why? Because when life in its infancy became capable of making food from sunlight, hydrogen, and carbon dioxide of the air, it did so. The carbon dioxide from the air was sucked up by organisms using raw sunlight to grow. Atmospheric carbon dioxide thus became grounded as parts of organic bodies; it entered into sugars, DNA, RNA, enzymes, structural proteins, hormones, neurotransmitters, bones, and other compounds crucial to living matter. Many of the carbon compounds of bodies that have died have been buried at sea and in the sediment. Since the early spreading of photosynthetic organisms over an anaerobic world, life has hoarded the organic valuables and kept them in the family. Later, bacterial descendants lodged in the leaves of plants pumped more carbon dioxide out of the air. Not much carbon is left in the atmosphere because it is stowed in the bodies of organisms and former organisms—mostly it is in limestone and shale rock, and in black coal and oil. Indeed, when we, at the brink of the 21st century, burn fossil fuels to provide our species with the energy to sustain civilization, we drain resources that were millions of years in the making. At the same time, we return carbon dioxide to the air making our atmosphere a bit more like the lifeless ones of Mars and Venus, with their surfeits of carbon dioxide.

▪ A THOUGHT EXPERIMENT ▪

The preceding was a fast tour through prehistory. We hope you now have a better idea of your origins. By looking at our chart, which is only a rough gauge of the relative dominance of organisms through geological time, you can see that life on Earth has been by and large a microbial phenomenon (Figure 2.10). Microbes invented all the most important biochemical and metabolic tricks, including fermentation, photosynthesis, and respiration. Some organic compounds, such as man-made plastic and drugs, "essential" plant oils like peppermint and eugenol, plant and fungal hallucinogens, and Amazonian snake venoms were not first synthesized in the microcosmic world. But these are the exceptions. When we insist that microorganisms are only germs we need to be rid of, or merely tools for genetic engineering of insulin or other substances of economic value, we divorce ourselves from our history. We dissemble our true nature.

Let us conduct a thought experiment. First, let's agree that the history of the microcosm represents about four fifths of that of the Earth. This will weight our perception so that we do not pay undue attention to recent developments. Now let's look at the evolution of life on the planet not as a motley, random collection of unrelated events, but as major episodes in the life cycle of a single planetary being. Life begins, eating food provided directly by the sun's ultraviolet synthesis of organic molecules. As damp patches of bacteria, life is challenged early on by deprivations: first by the disappearance of pre-made food, then by the decrease of hydrogen sulfide and lightweight hydrogen gas after the evolution of photosynthesis. Yet Earth

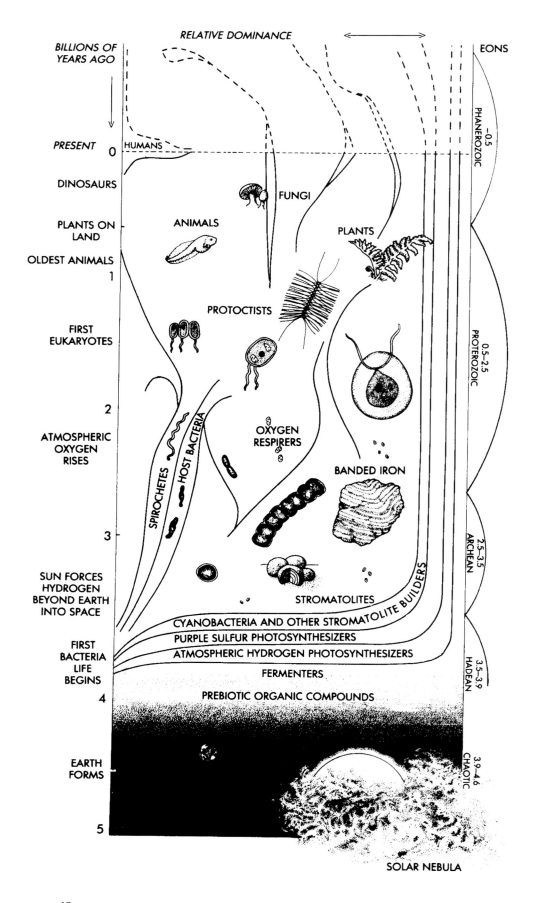

48

life—now purely microbial—responds brilliantly to the challenge. Confronted with dwindling supplies, bacteria evolve a new form of photosynthesis that drags out the precious hydrogen from a substance more prevalent on Earth than earth itself—water. However, the solution creates another, even greater problem: poisonous oxygen is produced in great abundance as a by-product of photosynthesis. In still another coup, life responds with oxygen-using—that is, oxygen-respiring—forms of bacterial life. From these varied, highly organized bacterial components, the energetic eukaryotes emerge. The eukaryotes diversify and become more complex, inhabiting the waters and the land. Under the influence of intense competition and eon-long wars between predators and prey, thought itself evolves. Smarter animals are better at capturing prey, obtaining food, and avoiding predators.

In our thought experiment one thing becomes clear: the mixture of danger and opportunity—the biological calculus that turns the worst crisis into the next form of survival—is a general theme of life. When life is perceived as a continuous phenomenon, the history of humans becomes a minor episode in the history of the biosphere. No matter how dearly we wish to separate ourselves from nature, we are stuck with it. Technological advances after World War II, for example, fall right into the ancient evolutionary pattern of responding to unparalleled dangers with unparalleled innovations. Human beings are not dominant over, but form a continuum with, nature. Isn't it clear that the evolution of thought and science, which so ably solve so many problems, led rapidly to the creation of so many others?

The rise of the intellect that made our feeble-bodied ancestors strong with spears and flaked stones in place of horns, teeth, massive bodies, and other congenital attributes of powerful predators is now proving to be as destabilizing as the release of oxygen into the atmosphere once was. Progress has not stopped with clubs and cave paintings, but has marched relentlessly on to computer-based ballistics and technologically unstoppable, submarine-based thermonuclear warheads. Where will it take us from here? Is "intelligence" uniquely doomed, a self-regulatory, interesting, but somewhat freakish phenomenon upon which nuclear war acts, when it crops up, as a controlling mechanism? Or is it rather something more in keeping with other pivotal changes in the biosphere, such as the origin of photosynthesis and microbes able to run on waste oxygen? If the world-changing, human-style technological intelligence is indeed part of the orderly progression of the developing biosphere, we have a biological basis for optimism. If the epical innovations of the past serve as a guide, then thought too—like water-using photosynthesis or oxygen respiration—may, after a destabilizing period of adjustments, infiltrate all corners of life and become grounded in the being of the biosphere. If the past serves as a guide, thought too will become integrated into global life—and lead to new, problematic forms beyond prediction.

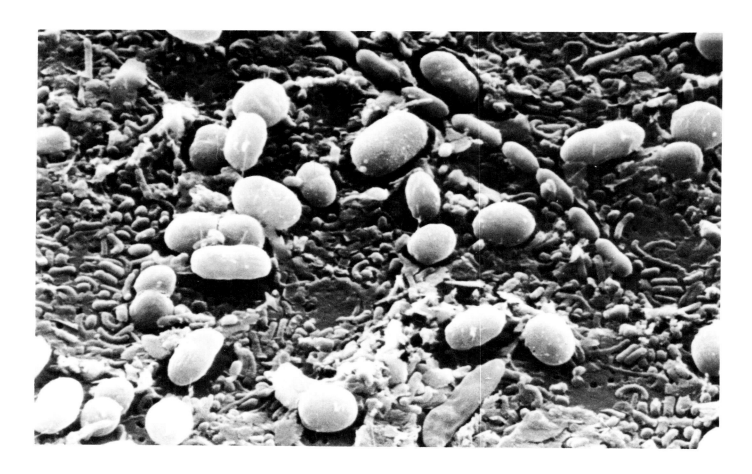

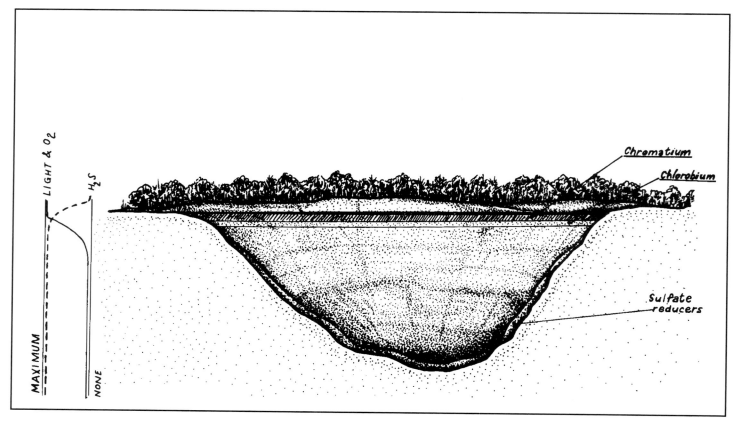

T·H·R·E·E

BIOSPHERE
AND
BIOTA

▪ MICROCOSMIC HABITATS ▪

Biologists tend to have difficulty isolating and growing pure strains of specific bacteria. Inevitably, other kinds of microbes get into the tiny habitat. It is the same with the planet as a whole. Microorganisms of the world unite because it is their nature to grow and interact, accommodating each other rather than going extinct. They form confederacies, recycle each other's by-products, and live and die with the nonliving materials around them in intimate or loose collectives. Certain habitats limit the growth of certain types of microbes. Each small and distinct place, such as a tide pool, the bark of an oak tree, or the feet of a heron, provides a special environmental opportunity for a specific group of interacting microbes.

The number of environments available to fungi, protoctists, and bacteria far surpasses the number of places in which plants and animals can live. This is not only because microbes are smaller, but because they are more versatile. Because bacteria reproduce faster and have been around longer than larger organisms, they have lived many more generations than animals or plants. Because each generation is in effect an evolutionary experiment, bacteria have been able to adapt to a wide range of different habitats. Some flourish in boiling hot springs and very hot acids. Others survive freezing temperatures, and some have even been collected with air filters from the stratosphere. By surviving intact in dehydrated forms, some bacteria keep a reserve copy of their genes in hard-walled spores that can begin to grow again when moisture is available. Microbial habitats include not only most of the surface area of the planet, extending down more than 10 miles into seawater and bottom and up over 6 miles into the air, but they also live on and within the bodies of larger organisms. Thus, Apollo astronaut microbes in the late '60s and early '70s completed their short life cycles on the moon. They reproduced 250,000 miles from the Earth's blue shores.

FIGURE 3.0
Profile of Lake Cisó, a sulfur-rich lake in northeast Spain. This lake surface is red because of the huge number of red phototrophic bacteria (*Chromatium*). The graphs to the left depict the light, oxygen, and hydrogen sulfide concentrations with depth. Bacteria from the *Chromatium* layer of Lake Cisó are shown in the scanning electron micrograph above. (Drawing by J. Steven Alexander, photo courtesy of Isabel Esteve and Ricardo Guerrero.)

51

Virtually wherever we look we find thriving hordes of associated microbes. The unparalleled growth potential of bacteria is such that, with appropriate quantities of food and water, a single bacterium would grow to the mass of a planet in several weeks. Such rapid expansion would be impossible for, say, people who require some 20 years instead of 20 minutes to reproduce.

Although there are nearly as many microhabitats as there are microbes, a couple of macrohabitats can be distinguished that pertain to our entire planet. The first of these is the surface area in which we and most organisms familiar to us live, the oxygenated environment inhabited by aerobes. Here free oxygen comprises some 20% of the atmosphere and exists in significant proportions, due to aeration and movement, in lakes, rivers, and oceans.

Below the familiar world rich in oxygen is an oxygen-poor world inhabited by anaerobes—organisms that do not use free atmospheric oxygen in their metabolism. Some are tolerant of oxygen, some are poisoned by the gas, and some use it facultatively, which is to say on a part-time basis. Modern prokaryotic anaerobes probably evolved from bacteria that appeared before the planet's surface became oxidized. Some, such as those that thrive in lush communities near thermal vents at the ocean's bottom, may never have experienced oxidizing environments.

One sort of habitat found on Earth today may recreate conditions of billions of years ago. These are the sulfur-rich waters and muds, collectively known as the sulfuretum. Sulfureta may be deep, narrow puddles a few centimeters across in evaporite flats, or they may be like the one in Lake Cisó in northeastern Spain (Figure 3.0). This sulfuretum is 11 meters deep and full of an abundance of sulfur- and sun-seeking, oxygen-avoiding microbes (see Figure 3.2). An absolute requirement of sulfureta is access to elemental sulfur, gypsum (calcium sulfate), or some other source of sulfur that can be turned into sulfide by microbes. Sulfureta are essentially bacterial territories. Oxygen and sulfide are extremely incompatible chemically, acting quickly to form sulfate compounds. In our oxygen-rich world, sulfureta are thus far less prevalent than they must have been in Archean times, 3.9 to 2.5 billion years ago.

The most productive members of sulfureta are the green and purple photosynthetic sulfur bacteria. These must be close enough to the surface of the water to live directly off the solar energy. Sometimes sulfur-tolerant blue-green bacteria also inhabit the surface of a sulfuretum. The sulfuretum is not a place for eukaryotes. Animals, plants, and fungi are excluded from this microhabitat because hydrogen sulfide poisons their respiratory systems.

Most protoctists are equally unable to inhabit the sulfureta. Yet a few specialized kinds have adapted and thrive in these smelly zones. Unicellular beings called ciliates (such as *Metopus contortus* and certain plagiopylid types) may be found in abundance, apparently counteracting or protected from the sulfur poison. Indeed, a close look at some of these ciliates reveals gas-mask equivalents, special features that allow them to overcome sulfurous fumes. Other ciliates have worked out arrangements with bacterial partners to mitigate the effects of the sulfur. A ciliate of the genus *Kentrophorus* has a bent, transparent body that harbors a "kitchen garden" of purple photosynthetic sulfur bacteria. Presumably the bacteria provide the ciliate with some sort of food.

52

Because they run on oxygen, mitochondria pose a great problem for eukaryotes wishing to adapt to the sulfur environment. Some ciliates have lost their normal-appearing mitochondria in what appears to be an attempt to adapt to the ancient, anaerobic conditions of the sulfuretum. The plagiopylids have weird striped bodies in their cells. Magnified by the electron microscope, these bodies look like alternating dark and light types of endosymbiotic bacteria. Other ciliates found here have so many surface bacteria that they seem to be wearing living coats on their outer layer. Apparently sulfuretum real estate has the advantage of proximity to bacterial food sources, but has required the evolution of special protection in the small number of eukaryotes that can survive in an environment still basically hostile to them.

All major habitats on the surface of the earth are composed of a superficial oxic, or oxygen-rich, layer where aerobes thrive, and an underlying anoxic, or oxygenless, zone populated by anaerobes. However, the ratio of oxic to anoxic portions in each habitat may vary greatly. To simplify matters a bit, we can carve up the planetary environment into five major regions: the coastal zones or seashores, the temperate forests, the deserts, the mountains, and the tropical forest regions.

In temperate regions, the seashore enjoys relatively mild weather and three types of shoreline prevail: rocky, sandy, and muddy. The temperate forest region, with its lakes and rivers, supports fertile, carbon-rich soil. Deserts and high mountain habitats favor organisms that store water or find it in novel ways. The tropical forest habitat is rife with lush vegetation and thousands of undescribed species that await discovery. The Earth, even with its local diversity and recognizable communities, is really one giant habitat, the biosphere. Every member of the biosphere, from the low-oxygen vent communities in the ocean's abyss to the microbes flavoring wine sipped by an airline passenger, ultimately interact and contribute to it.

During most of the history of life on Earth, microbes have been constrained by limited supplies of food and water. Natural limitations and dangers are called *selection pressures*, a term coined by the British biologist Julian Huxley and used to express those forces that affect evolution by permitting or restricting population growth. Selection pressures coupled with the incessant tendency of microbes to grow have induced extremely complex microbial relations. Present survivors have all competed—and cooperated— with each other's ancestors in unimaginable crowds equivalent, say, to eight billion people crammed, jammed, and nestled into downtown New York. Such density induces selection pressures of its own.

Selection pressures and short generation times have helped produce microbes that not only live in each other's wastes but recycle these by-products, making them available for other microbes and some plants. The extent of interliving and mutual microhabitats includes complex communities of bacteria living within animals as tiny as insects. We saw microbes living inside other microbes in the dense communities within the Florida termite hindgut. Besides animal innards there are many other fascinating structures, such as microbial mats, that are formed by compact interwoven masses of interdependent microbes. The illustrations show close-ups of the surface of microbially rich seaside mud. On top lives a photosynthetic mass of blue-green bacteria (Figure 3.1); beneath it, an anaerobic layer of consumer

FIGURE 3.1

The top layers of a microbial mat community. This drawing shows a cutaway view of the cyano-bacteria and other members of this complex community that inhabit the surface of rich seaside muds. (Drawing by Christie Lyons.)

FIGURE 3.1

The top layers of a microbial mat community. This drawing shows a cutaway view of the cyano-bacteria and other members of this complex community that inhabit the surface of rich seaside muds. (Drawing by Christie Lyons.)

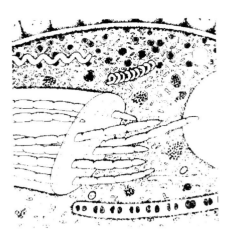

FIGURE 3.2

Anaerobic phototrophic bacteria form the communities in oxygen-poor areas such as the surface layers of Lake Cisó when it turns red. The largest cells in this drawing are *Chromatium* sp. Also included are *Chlorobium, Vampirococcus, Daptobacter,* and some spirilla. (Drawing by Christie Lyons.)

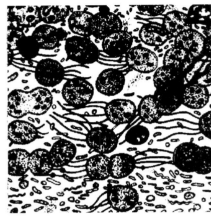

organisms poisoned by oxygen (Figure 3.2). Millions of millions of microorganisms are so packed together that under the electron microscope the layers look like tissue. Like the sulfureta, areas of the world covered with microbial mats can be studied to give us insight into the Earth's ancient environment. Most forms of microbial interaction are still waiting to be discovered.

▪ MICROBIAL MATS AND STROMATOLITES ▪

Microbial mats are sticky, textilelike expanses of mud, usually found in intertidal regions where the strong wave motion of the open ocean is minimized by a barrier of dunes or rocks. Mats are often spectacularly colored purple and green as a result of massive populations of photosynthetic bacteria. Microbial mats probably represent the way much of the life on Earth looked in the first billion years or so.

Today microbial mats can be found in many places, including along the coasts of Baja California, Mexico and southeastern United States, and in the Caribbean islands. Mats usually flourish in areas that are too salty or stagnant for larger organisms. The crucial members of an intertidal microbial mat community are always blue-green bacteria. We have met this venerable group of photosynthetic bacteria before when they accomplished the energy

54

revolution. Although different types of blue-green bacteria have different metabolisms, they all are capable of using water during photosynthesis and most produce oxygen when they are in the light. They split hydrogen molecules from water and hook them to carbon dioxide molecules to make the organic compounds of their bodies. Some blue-greens can switch to other substances, such as hydrogen sulfide, to supply their hydrogen when conditions warrant it. If blue-greens use hydrogen sulfide as a source of hydrogen, they deposit sulfur rather than give off oxygen. The blue-green *Oscillatoria limnetica* and some *Phormidium* are chameleon organisms that can switch their metabolisms according to the conditions found in their microhabitat.

Blue-green bacteria spread along the surface of a microbial mat and crawl out of tough polysaccharide sheaths (such as the one shown cut-away in Figure 3.1) as they move upward in search of the sun. Because these bacteria, like farms, are prodigious makers of food, many "consumer" bacteria that are not poisoned by oxygen inhabit the mats as well. Below the oxygen-tolerant layer are other bacteria, such as purple anaerobes, that are in the precarious situation of requiring light yet being poisoned by the oxygen produced by blue-green bacteria. Together the mat microbes form layered communities that depend entirely on photosynthesis for their sustenance. Bacterial environments remain in many parts of the world today, but they have been contaminated by larger, nucleated microbes—the protists. Nematodes and brine shrimp are among the few animals capable of living in the microbial mat environment.

The blue-green bacterial layer grows upward, leaving behind sheaths that trap and bind sand. Several species of blue-greens precipitate grains of the hard chalky material, calcium carbonate. Growth may take hundreds of years, but microbial mats will form solid structures if conditions are right. These layered rocks, generally made of calcium carbonate (limestone), are called stromatolites. They represent the petrified remnants of massive populations of bacteria living in layers—a sort of ghost megalopolis of abandoned skyscrapers. Stromatolites are the first large, complex, and unmistakable evidence of life in the rock record (Figure 3.3). As such, they hold a particular fascination for naturalists, amateur or professional. While some fossil stromatolites have been dated close to three and a half billion years old, others—limestone rocks coated with live blue-green bacteria—are still growing today.

Limestone stromatolites have been found in remote intertidal areas from the Bahamas to the western coast of Australia. One type, originally called cryptozoans, meaning "hidden animal"—a two-kingdom appellation—can be found cropping up near Saratoga, in upstate New York. Only in the late 1950s, with the rise of studies of early life, were stromatolites recognized as bacterial structures. Some are domed mounds that look a bit like the top of a bowler hat. Many stromatolites contain thousands of layers, representing centuries of growth of sun-seeking microbial communities (Figure 3.4).

While modern, still-growing stromatolites are relatively scarce, fossilized ones suggest they were much more abundant before the evolution of nucleated cells and, later, of animals and plants. A realistic scenario attributes the fall in production of these rocky mounds to the evolution of bacteriovore protoctists and the earliest grazing animals. Such organisms, more than half a billion years ago, must have considered the microbial mats delectable and devoured

FIGURE 3.3

Closely-spaced stromatolite mounds from the Proterozoic Kuuvik Formation, Kilohigok Basin, Victoria Island, Northwest Territories, Canada. These domed stromatolites are about two billion years old and over two meters high (the bar scale is divided in 30 cm units). (Photo courtesy of Fred Campbell.)

FIGURE 3.4

Structures like these formed by layered microbial communities are called microbial mats. Certain microbial mats that bind particles of sediment (such as silica sand or bits of carbonate) are the precursors to rocks called stromatolites. (Photo courtesy of Robert Horodyski.)

them faster than they could grow back or harden into stromatolites. Furthermore, when the rugged photosynthetic protists and algae evolved they began to compete with the blue-greens, moving into provinces where only bacteria had lived before.

While it is widely accepted that stromatolites are fossilized bacterial communities, it is quite possible that other rocks also result from bacterial activities. Beside banded iron formations, deposits of calcite, gypsum, barite, opal, and other minerals may be helped along by microbes, which are known to extract calcium carbonate, sulfates, barium, silica, and other mineral compounds from the solutions in which they live. Figure 3.5 clearly depicts *Bacillis* bacteria depositing the mineral manganese dioxide around its cells.

But of all the mineral accumulations that may owe their existence to microbes, none is more exciting to the imagination than gold. In the most famous mining region of the world, the Witwatersrand of South Africa, gold is found in mines by following a seam of black carbon, known as the carbon leader. In addition to containing deposits of gold, the carbon leader leads to layers of fool's gold (iron pyrite) and to uranium ore. The coexistence of carbon—a crucial constituent of DNA and proteins—with the precious gold suggests that prokaryotes may have been involved in the layered deposition of the metal gold just as they were in the construction of stromatolites. Certain modern bacteria can bring gold out of solution as a consequence of their metabolism and even concentrate microscopic gold grains into visible deposits. Furthermore, the carbon leader contains spherical and filamentous microfossils. Whatever their reasons, or rather, selection pressures, the first goldmongers and collectors may well not have been human at all—but bacteria!

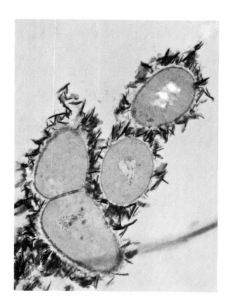

FIGURE 3.5
Bacillus bacteria surrounded by manganese deposits. (Transmission electron micrograph courtesy of John F. Stolz.)

57

MODES OF NUTRITION

NAME OF MODE / EXAMPLE	SOURCE OF ENERGY	SOURCE OF CARBON
AUTOTROPHY		
plants, algae, cyanobacteria	light	atmospheric CO_2
sulfide−, methane− and ammonia− oxidizing bacteria	inorganic compounds (H_2S, CH_4 and NH_3)	atmospheric CO_2
HETEROTROPHY		
fungi		
protoctista (ciliates, mastigotes, slime nets, etc.)	organic compounds (C,H,N,O)	organic compounds (C,H,N,O)
animals (mollusks, hydras, fish)		

F·O·U·R

MAKING A LIVING
AND
STAYING ALIVE

■ NUTRITION ■

The experience that heat flows from a higher to a lower temperature but not in the reverse direction without a continuous input of energy has staggering implications. It touches upon the irreversible nature of time. It suggests that the universe is "running down," becoming more and more molecularly blasé, inert, and uniform. It almost seems a law of nature. Yet set against this tendency stands life itself. Compared to other planets, infrared photographs of the surface of the earth reveal a great patchiness in the distribution of heat and cold. And, while our sun has grown some 40% more luminous over the past 4.6 billion years, the earth has remained at tolerable temperatures for life all the while. We know this from the fossil record, which stretches back well over three billion years.

It is agreed that the earth must have cooled by a combination of a steady reduction of radioactivity inside it and a lowering of the quantity of carbon dioxide in the atmosphere. This could have been sheer luck. But it also is the nature of life to covet pattern, to remake its environment, to become more organized and complex. Darwin forwarded the idea that evolution, like laissez-faire economics, always works toward the good and perfection of the individual. Although this idea is debatable, in surveying the long history of life we witness trends of increasing size, complexity, even self-awareness. Does life, then, uphold the pessimistic 19th-century notion of "heat death" based on the experience of heat's one-way flow: that the evolution of energy and matter, if not information, is inexorably headed toward a final state of maximum disorganization?

Not necessarily. Thermodynamics, from the Greek for "heat movement," is the study of the group behavior of massive numbers of particles. Living beings are so far away from the idealized state of thermodynamic equilibrium that they have entered the realm of thermodynamic disequilibrium structures.

FIGURE 4.0
Comparison of modes of nutrition.
(Chart drawn by J. Steven
Alexander.)

59

These structures are no longer treated by scientists as temporary disturbances of equilibrium, but as pockets of matter and energy that are racing away from equilibrium. The biosphere as a whole thus expands, becoming more intricate and complex through time, somehow ridding itself of entropy—energy in the form of heat, which cannot be used to enhance organization. This idea comes in part from the new field of disequilibrium thermodynamics, pioneered by Ilya Prigogine, and may be seen as a sort of abstract mathematical apology for physical principles that were prematurely applied to biology.

The science of thermodynamics itself arose in the 19th century along with the development of the steam engine and other machines. It became apparent that thermal engines can never be totally efficient. Automobile engines and steam engines never completely turn heat into work; much is inevitably lost as thermal pollution to the environment. These industrial ideas were not meant to be applied to the more complex and intricate operations of organisms. Living beings are not describable by direct comparison to mechanically engineered devices. The famous mathematician and pioneer computer scientist John von Neumann suggested that complexity, after it reaches a certain point, becomes self-perpetuating. Even when viewed as no more than biochemical machines, organisms display a complexity that catapults them into a region where standard physicochemical analyses of inanimate objects are inadequate.

The sum of organisms on Earth—the biota—has become both more diverse and more complex over the last three billion years. Like a steamship winding its way up the Mississippi River or the construction of the Empire State Building, such evolution depends on a source of energy. But what is done with this energy, which ultimately derives from the Sun, and how it is done are quite fantastic. Whether we look at the biota of the past or of the present, we can distinguish three basic modes of nutrition or of gathering energy (Figure 4.0). Each method gives organisms the energy to increase their order, to maintain their pattern of identity, to grow and become complex regardless of the second law of thermodynamics, which says that the entropy of an isolated system can only increase until the system has reached its thermodynamic equilibrium. Organisms that give off heat and waste and that use up lots of energy just keeping themselves intact do not defy the second law, they just transcend it.

Paradoxically, the first mode of nutrition to be used by organisms was probably heterotrophy, which means "living off others." This is paradoxical because, of course, there were no "others" when life first evolved from a nonliving environment. But the Sun's ultraviolet radiation synthesized and broke apart the carbon and hydrogen compounds that were to become life. When life evolved, the nonliving leftovers acted as pre-made nutrients and were absorbed by bacteria. Today all animals and fungi, and most protoctists and bacteria, are heterotrophs. Even some parasitic plants, having lost their ability to photosynthesize, are heterotrophic: unable to make their own food, they "eat others."

The first heterotrophs indirectly required sunlight to make their food through the energizing action of ultraviolet radiation on gases and inorganic mixtures in liquids that, in turn, yielded carbon-containing chemicals. Modern heterotrophs also ultimately depend on the Sun because they eat plants and other organisms that photosynthesize. While some organisms eat beings that make their own food, all life ultimately depends on the productivity of photosynthesis, which is essentially a bacterial process.

The next mode of nutrition to appear chronologically comes first in terms of importance. Photoautotrophy refers to the process whereby organisms generate their own energy and make their own food from sunlight and inorganic compounds, usually atmospheric carbon dioxide—in other words, by photosynthesis. The photoautotrophs were to the heterotrophs what herbivores are to carnivores. Their mode of nutrition really runs the entire biosphere. Plants are photoautotrophs that take carbon dioxide from the air and, combining it with hydrogen from water, make carbon-hydrogen compounds such as wood, starch, protein, and sugar. (They also put carbon dioxide back into the air.) Purple sulfur bacteria are photoautotrophs that take carbon dioxide from the air and combine it with hydrogen or hydrogen sulfide gases seeping from the ground to make carbon compounds, allowing them—and their predators—to grow and reproduce. Purple nonsulfur bacteria acquire carbon in the form of carbon dioxide from the air, but they take the hydrogen from neither sulfur compounds nor water but from small organic molecules such as lactate, pyruvate, or ethanol. Together photosynthetic organisms deplete the atmosphere of an estimated 250 billion tons of carbon annually as they weave elements into the organization called life.

Some of the carbon that is returned to the air is removed from it by another kind of autotrophy, which forms the third basic mode of nutrition. Chemoautotrophy depends neither on sunlight nor food in the conventional sense. Sometimes called chemolithotrophs or, in this book, gas eaters, chemoautotrophs do not require pre-made organic nutrients; they are truly independent. Chemoautotrophs make more of themselves by living off carbon dioxide from the air, salts, water, and a nonbiological source of energy, such as hydrogen sulfide, ammonia, or methane gases. The gas eaters thus are dependent on neither light nor others but on inorganic chemical reactions for their energy. Elemental carbon, nitrogen, and sulfur are always hot commodities because of their utility and scarcity. Necessary for the growth of all organisms, these elements are kept in circulation by many processes, including the activities of gas eaters.

Chemoautotrophy as a mode of nutrition is unknown in plants, animals, fungi, and protists—it is strictly a bacterial phenomenon. Although photoautotrophy is by far the major system of productive nutrition on Earth, in recent years underwater ecosystems based on chemoautotrophic food producers have been discovered. These communities include the giant red tube worms, called vestimentiferans, as well as specialized clams and crustaceans. The deep-sea vent ecosystems depend on bacteria that oxidize reduced chemicals as a source of energy—mostly hydrogen sulfide but sometimes methane—that seep up from ocean bottom thermal vents. Like algae and plants, these bacteria use the carbon dioxide in the water to make their food. This type of chemoautotrophy may have been prevalent on the early Earth when the lack of predation, the presence of hydrogen-rich gas from the interior of the Earth, hot shallow seas, shorter days and nights, influxes of meteorites, violent storms, and general prebiotic mixing created a chemical environment favorable to bacteria that could feed and grow off gases and energy derived from inorganic chemical reactions.

The discovery in 1977 of these dark gardens was a great shock to scientists. The oceanographers felt compelled to bring biologists to visit the sites and explain: How could life survive without the energy of the Sun to directly power the photosynthetic show? We now know life can thrive on chemical energy coming through cracks in the ocean floor. Life probably has survived in special dark oases like the deep sea vents for millions of years. But think twice. Oxygen is needed to oxidize the hydrogen-rich gases seeping from the Earth's energy-rich interior, even in the cold abyss. And this free oxygen ultimately comes from the surface of the Earth where it is produced by cyanobacteria, algae, and plants. Thus even dark gardens indirectly depend on the innovations of photoautotrophs and our life-giving star—the Sun.

▪ MOVING AROUND ▪

Almost as impressive as the metabolic diversity of microbes is their mobility. First let us look at movement in prokaryotes; then we will progress to the more complex movements of cells with nuclei. The first form of movement undoubtedly was passive: simply floating with water or air currents. But even an entirely passive bacterium moves by expansion as the cell grows. Morphogenetic movement, the movement that accompanies growth, is a property of all life. Buoyed by surface tension and swept along by currents, the first bacteria grew and spread. The blue-green bacteria and green photosynthetic anaerobes ultimately moved to almost every corner of the Earth in this slow but steady fashion.

Beside morphogenetic movement, two kinds of active movement have been described by bacteriologists: gliding and swimming. Gliding, a simple but eerie form of microbial locomotion, is relatively slow. Certain bacteria encased in polysaccharide slime or stuck in a layer of water on a glass slide unmistakably glide along surfaces. To keep going they never lose contact with the glass or rock or twig. Although electron micrographs show tiny intracellular fibrils in some gliding bacteria, their means of mobility is still largely a mystery.

Swimming bacteria literally invented the wheel three billion years ago. The motility structures of swimming bacteria are known as flagella (singular: flagellum), and are composed of one or another of a class of characteristic proteins, called flagellins. The flagella, which tend to be organized into bundles, do not move; they are simply whips or tails attached to the rotating parts, or basal wheels, inside the cell. These basal wheels turn by internal electricity, a flow of hydrogen ions (H^+) called a proton motive force. A difference in electrical potential may also determine the swimming direction of the microbes. Bacteria "run," swimming directly toward a food source, light, magnetic field, or whatever interests them. They also periodically "twiddle," moving about in a more random fashion. To avoid unpleasant conditions, such as too much acid or too high temperatures, the rate of "twiddling" increases. After a "twiddle," bacteria "run" again in one direction. Many "twiddles" tend to keep bacteria from running toward something. "Twiddles" occur when bacterial flagella rotate in their reverse direction and then the flagella bundles splay (Figure 4.1).

Bacterial swimming is seen in spirochetes and spirilli—both slender, helically shaped bacteria that dart about due to flagellar rotation. In spirochetes, however, the flagella are internalized beneath the outer layer of membrane of their cell walls. The result is the rapid corkscrewlike move-

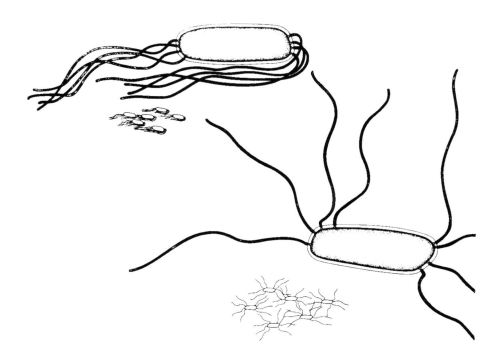

FIGURE 4.1
Bacterial motion. Bacteria move by alternating between "running," directed swimming in which flagella rotate in a coordinated fashion (top), and "twiddling," a random motion in which flagella are splayed out (bottom). (Drawing by Christie Lyons.)

ments in either "head" or "tail" directions that make spirochetes the foremost speedsters in many viscous microhabitats. In the termite hindgut habitat, for example, crowded populations of spirochetes swim in unison, the wavelengths of their wriggling bodies synchronizing into a united movement like uniformed marching-band members at a football halftime show. Spirilli, bacilli, and vibrio bacteria have propulsive flagella that protrude beyond the outer layer of membrane of their cell wall (Figure 4.2).

Although the growth of all cells, bacteria included, generates locomotion in the form of morphogenetic movement, some organisms have used morphogenetic movement in the evolution of certain novelties. Gliding bacteria, for example, (called myxobacteria, meaning "slime bacteria" in Greek) join to form "fruiting bodies," "trees," and "rosettes." And, as we have seen, the inert layered limestones called stromatolites are formed by blue-green bacteria shedding their sheaths to produce new layers. Indeed, the rocky stromatolite domes are actually the stationary result of microbes in motion.

Eukaryotes—plants, animals, fungi, and protoctists—are entirely distinguishable from prokaryotes on the basis of cellular motion as well as their nucleation. Eukaryotes, like prokaryotes, not only move as entire cells or organisms composed of cells but—unlike prokaryotes—also always display an intracellular system of movement. The constant streaming and transport of materials seen in eukaryotic cells is called cyclosis. This incessant intracellular movement occurs within the cytoplasm, the fluid portion of the cell between the cell and nuclear membranes. In many eukaryotes, the streaming occurs along internal trackways seen with the electron microscope to be composed of fine tubes, called microtubules.

Eukaryotic cells also move by means of external, tail-like organelles. Although these still are often referred to by the misleading name of "flagella," they lack the flagellin protein and differ in every detail from bacterial flagella

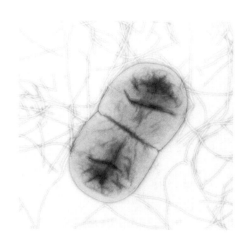

FIGURE 4.2
A *Bacillus* dividing. It is surrounded by its external flagella (which fell off in preparation of this specimen for electron microscopy.)

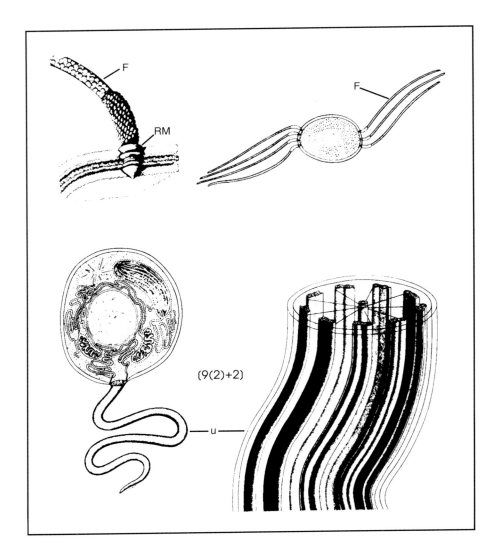

FIGURE 4.3

Prokaryotic cells (top right) move by flagella (top left) composed of flagellin protein. The flagella (F) spin on a rotary motor (RM). Eukaryotic cells (bottom left) move by means of undulipodia (u) (bottom right), "waving feet" composed of tubulin microtubules arranged in a characteristic (9(2)+2) pattern. Many other proteins, including some called MAPS (microtubule-associated proteins), are present. The sliding of these microtubules along one another causes the undulipodium to wave. (Drawing by Christie Lyons.)

(Figure 4.3). Because they are not at all like the far smaller bacterial flagella, and because cilia, sperm tails, and all the so-called eukaryotic "flagella" are identical, it is better to refer to them by a clear, unambiguous name. Several scientists writing early in this century made the same suggestion of uniting the identical motion organelles of eukaryotes under one common term. The word "undulipodia" has been used for this purpose and we will also use this term.

Undulipodia (literally "waving feet") are about a quarter of a micron in diameter; most vary from 1 to about 20 microns in length. Undulipodia display a special internal pattern that identifies their common origin as surely as a fingerprint does a suspect. This pattern, which we witnessed with the transmission electron microscope in our journey into a termite cell (see Figure 2.0) is referred to as the [9(2)+2] structure because it consists of nine fused pairs of microtubules surrounding a central pair. This pattern is found in the cells of many eukaryotic organisms: at the base of the cilium of a ciliate; in the undulipodia lining our tracheas; in the swimming, sperm cells of a moss plant. That such a specific structure is found so often in so many different organisms adds another support to Darwin's insight that we have all descended with modification—in this case from cells.

The process of eukaryotic cell division, known as mitosis, repeatedly has been observed to be directly connected with the process of undulipodia formation. In mitosis, tiny long fibers seen with the light microscope form what is called the mitotic spindle. Under the electron microscope, the filaments prove to be bundles of microtubules of the same diameter as the microtubules those of undulipodia. Furthermore, both mitotic microtubules and those of undulipodia are composed of tubulin proteins. Mitotic spindle microtubules emerge from the ends of the cell and the surfaces of the chromosomes and carry off chromosomes in some unknown way to offspring cells. Although cells in mitotic division can afterward grow undulipodia, once animal and plant cells have formed undulipodia they never can divide again. Some of the earliest collectives of nucleated cells may have consisted of groups of cells divided into "movers" and "reproducers." By dividing the labor into mitotic reproducers and undulipodiated movers, our microbial ancestors may have been able to swim faster and gain a selective advantage over other life forms.

▪ MOVEMENT INSIDE PROTOCTISTS AND FUNGI ▪

Although prokaryotes lack internal movement entirely, microbial eukaryotes, such as protists and fungi, have several kinds. Not only does cyclosis occur but there also is motion as microtubules slide past each other in undulipodial movement. The bending and straightening of microtubules leads to the waving and swimming action of undulipodia. Nerve cells and the spikelike extremities of some protists display another kind of movement: the movement of particles along the surfaces of microtubules that constitute the internal trackways of the cell. All of these movements are relatively quick compared to the microtubule-based morphogenetic movements of mitosis or the growing out of undulipodia.

Cyclosis is involved in processes familiar to biology students: exocytosis, the transport of secretions and other materials outside a cell, and endo-, pino- and phagocytosis (Figure 4.4)—the transport of particles into the cell. Although no one is sure, all these cell "-oses" are most likely based on a protein system much like the one involved in animal muscles. The two most important proteins of the muscles, and perhaps of cyclosis, are actin and myosin. Together they make submicroscopic fibers that reversibly shorten in the presence of calcium ions. Microtubule morphogenetic movement, on the other hand, does not involve actin or myosin but rather depends on tubulin proteins coming together to form new microtubules. Movement of particles along the surface of formed microtubules represents a way for protoctists to circulate their food through their star-shaped or folded bodies. Particles of prey, mitochondria, granules of different colored pigments, all move along microtubules. It is possible that filaments of actin proteins on the surfaces of microtubules are involved in these movements.

Some protists, like *Stentor,* have still other, unique types of internal cell movement. *Stentor* can contract so quickly when threatened that even the highest speed films have difficulty tracking the organism's motion—it probably takes *Stentor* 10 millionths of a second to make the movement. The contraction of the *Stentor* body is due to M fibers, the composition of which

65

FIGURE 4.4

Types of cytosis (cell movement):
(a) endocytosis (cell eating a
piece of another cell in a tissue),
(b) exocytosis (cellular excretion),
(c) pinocytosis ("cell drinking"),
and (d) phagocytosis ("cell
eating"). (Drawing by J. Steven
Alexander.)

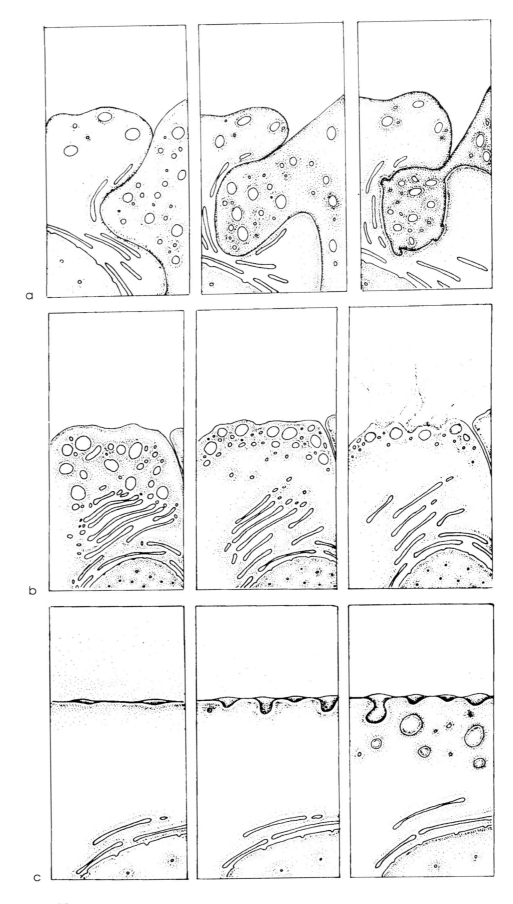

a

b

c

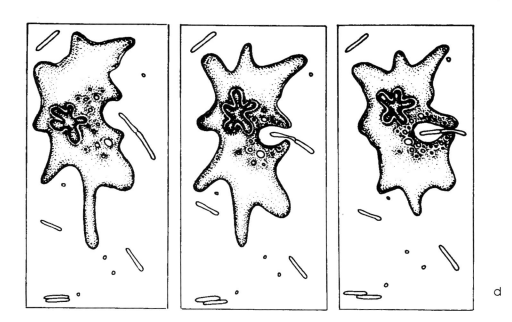

d

is not known. When stimulated, these fibers shorten dramatically, causing the protist to reduce to one seventh its former length. In summary, intracellular motion in eukaryotes is complex, strange, and difficult to study. Prokaryotic intracellular motility, on the other hand, doesn't even exist.

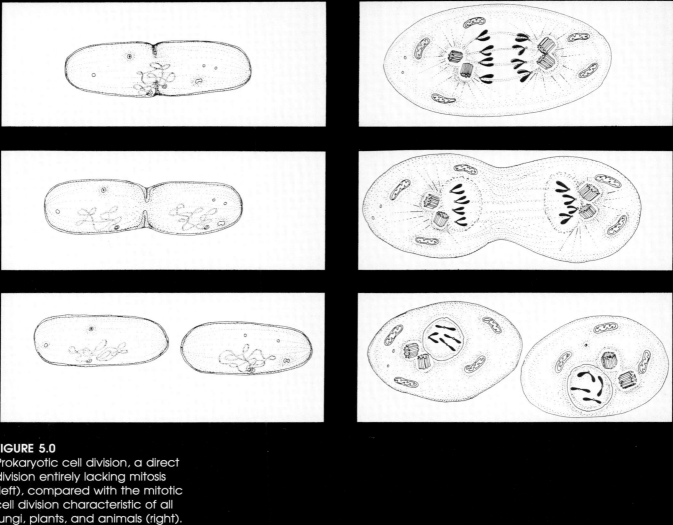

FIGURE 5.0
Prokaryotic cell division, a direct
division entirely lacking mitosis
(left), compared with the mitotic
cell division characteristic of all
fungi, plants, and animals (right).

GROWTH, MATURATION, AND SEX

▪ MULTIPLICATION BY DIVISION ▪

The differences between prokaryotic and eukaryotic cells extends to cell walls. Both types of life have semipermeable cell membranes, permitting the passage of selected chemicals. But bacterial cell walls are composed of glycopeptides (sugar molecules linked to tiny proteins) while eukaryotic cell walls are made of longer carbohydrates, either cellulose or chitin. Eukaryotic cells also evolved the ability to control the amount of calcium coming into the cell and to use this calcium in forming structures both inside and outside the cell. The results are hard parts, such as bones (calcium phosphate) and sea shells (calcium carbonate).

Prokaryotes and eukaryotes also divide in fundamentally different ways (Figure 5.0). Bacteria reproduce more quickly, directly, and less visibly than eukaryotes. New cell wall material is laid down internally and the cell simply splits apart. This direct division of bacterial cells is called binary fission. Eukaryotic cells, however, divide by mitosis or, occasionally, by meiosis. Both are complex processes that usually involve chromosome replication and deployment by the mitotic spindle and, in the case of meiosis, an exact halving of the genetic material.

Meiosis, known as "reduction division," rarely occurs. In this method of cell division the chromosomes are not duplicated prior to division and thus only one half the chromosomes carrying DNA go to each offspring cell. Such division necessarily occurs prior to or during the formation of the sperm and egg cells of plants and animals. These come together in fertilization, reforming a cell with a nucleus containing the full quantity of DNA and chromosomes.

The chemical details of growth and cell differentiation remain unsolved mysteries in modern biology. Though the chemical nature of replication is understood for the DNA of bacterial cells, deducing the development of animals from mathematical or biochemical rules is beyond present capabilities. The idea that individual eukaryotic cells are themselves the evolved outcome of bacterial communities does not solve the problem. To explain the development of an individual plant or animal, its bacterial constituents, as well as the differentiation of the fertilized egg cell, must be taken into account. Nonetheless, looking at larger creatures as ordered collections of symbiotic microbes ultimately should help in an understanding of embryonic growth and differentiation.

▪ MULTICELLULAR AND SINGLE-CELL SEX ▪

Many abstract ways of looking at the problem of eukaryotic cell differentiation have been proposed, among them the idea of "chreods," suggested by the British biologist Conrad Waddington. Ever since Haeckel suggested that "ontogeny recapitulates phylogeny"—in other words, that individual animals retrace the evolution of their species in the egg or womb—biologists have pondered embryonic patterns. Although Haeckel's declaration that the unborn progress through ancestral stages is no longer accepted at face value, the strong patterns of embryonic development suggest that there are evolutionarily beaten paths. Waddington in the 1950s called these paths "chreods," from the Greek *chreos,* meaning debt or obligation, and compared them to a ball bearing or marble rolling over a series of hills and valleys: the past route and momentum of the rolling ball limit and influence its direction without determining its exact path. For example, if one cell of a sea urchin is destroyed after the fertilized sea urchin egg has divided for the first time into two, the other cell will grow not into half a sea urchin but a whole one. This sort of sensible regulation—the molecular basis of which remains unknown—was what Waddington was referring to when he named the paths "chreods."

Like bacterial dots on a biologist's petri dish, large eukaryotic organisms, such as people, dogs, and oak trees, are clones—multicellular populations derived from individual cells. A major mystery of large creatures is that the supposedly identical copies of the original cell differentiate. The cells separate into special camps, dividing the labor into bark cells and leaf cells, or into muscle cells, intestine cells, kidney cells, bone or brain or uterus cells. However, the genes—as DNA bound to protein in the form of chromosomes—in the nucleus of any given cell of a plant or animal are assumed always to be the same in all its cells. This paradox represents the crux of the biological problem of differentiation: assuming, as everyone does, that the genes are responsible for making proteins, and thus for determining the shape and function of the cell, how can they all be the same in different cells?

Traditional theory holds that genes are turned on and off by genetic "switches," thus making the same set of genes produce different structures. The plant pathologist Rupert Sheldrake has suggested that a nonmaterial agency—that is, a nonenergetic but empirical force—that he has called a "morphogenetic field" acts in concert with genes to account for the differentiation of dividing cells into distinct tissues and organs. Such invisible fields would pull potentially similar cells into separate structures somewhat in the manner that invisible magnetic fields of a magnet pull iron filings into a discrete design. However, evidence of such fields has not yet been convincingly shown.

70

It also is possible that the widespread generalization that genes are the same in every cell is far from correct. Red blood cells, for instance, contain no nuclei. In a similar manner, it is possible that despite the presence of 46 chromosomes in virtually every cell of the human body the genetic constituents differ. Indeed, it has been shown in tissue cultures of tomato and tobacco plant cells that the cell division process of mitosis regularly gives rise to nonidentical offspring cells.

One approach to differentiation is to look at the simplest cases. Fungi form threads called hyphae and these threads join together into molds or mushrooms. Gliding bacteria differentiate into multicellular "trees" (Figure 5.1). Thanks to gene manipulation techniques, called genetic engineering, as well as the ease with which some fungi, such as yeasts, mate in the laboratory, some cases of cell differentiation are being worked out.

For example, two sexes in the types of yeast used to make beer have been intensely studied. Since they look alike, or rather have no obvious sexual features that allow them to be distinguished, the sexes are called "**a**" and "α". The only rule is that "**a**" makes a microbially sexy substance which attracts mating type "α" and therefore "**a**" cells must mate with "α." No "**a**" cells can mate with other "**a**" cells, nor can "α" mate with "α." Since mitosis is an equal process, all division products of an "**a**" parent should be "**a**" offspring. They aren't. Yeast collected from nature are usually "homothallic." Homothallic strains can develop more than one type of mating type from a single cell. For example, in homothallic yeast, an "**a**" parent regularly gives rise to some "α" offspring and, if such a parent is placed by itself in a microhabitat with food deficiencies known to induce mating (called "mating medium"), sex takes over and matings occur in great quantity. But on close inspection, they are not really homosexual matings because the cells, although coming from mitosis, have changed sex during the course of growth.

Suffice it to say that normal "**a**" cells have copies of "α" genes in an inactive state and "α" cells normally keep copies of "**a**" genes in an equally inactive state. Furthermore both mating types have a whole repertoire of gene splicing and recombining tools: DNA copying enzymes, transposase enzymes, and ligating (tying up) enzymes, among others. It seems to be a part of the normal differentiation in homothallic strains to change sexes on a regular basis by splicing the mating type gene, whether it be "**a**" or "α," out of the linear order of the DNA and placing it next to a "switch on" gene that determines whether or not the gene next to it will become active. If an "**a**" gene is next to the "switch on" gene, the cells find themselves very willing to mate with "α." Even in simple yeasts—really just tiny nucleated spheres— sex via differentiation into two types of cells already is a complex affair. It involves many genes, the production by each of the sexes of chemical compounds leading to sexual attraction and regularized changes in the linear order and activity of DNA.

FIGURE 5.1
Stigmatella, a colonial bacterium that forms "trees." Each packet at the end of the branch is full of billions of gliding bacteria. (Drawing by J. Steven Alexander.)

▪ STRATEGIES OF SURVIVAL: SPORES AND CYSTS ▪

Although sophistication is not usually associated with microbial life, this is only a habit of thinking. Even complex animals made of billions of cells, such as ourselves, represent specific answers to problems of microbial survival. We are bacterial crowd behavior, community

71

response patterns, strategies responding to stringent internal and external conditions imposed from within and without for many millions of years. But, since our genes stretch back in time in an unbroken line to the first life forms—just as the genes of mushrooms, bacteria, and rabbits do—it is questionable to assume that we and our domesticated mammals are higher forms of life. Certainly more recent, in some ways more complex, our intelligence is, nonetheless, simply another successful strategy for survival. But if enormous increases in population size do, as some biologists believe, immediately precede extinctions, human beings may not be long in this world. In any case, we should be wary of the hidden assumption that humanity is equivalent to evolutionary superiority.

This is especially true if our species continues to eliminate other species in its "progress." Ecosystems with great numbers of different organisms are biologically the most stable—they create great quantities of reduced carbon compounds usable by all life, and contain a natural reserve of diversity that can be tapped during an unpredicted environmental crisis. From the vantage point of global chemistry, some of the most valuable ecosystems on Earth are marshes, seaside mats, and other inconspicuous areas dominated by microbes. Such areas are veritable treasure troves of chemical and metabolic diversity.

Some of the cleverest survival strategies belong to bacteria. Two billion years ago there were no birds or leaves or seedlings in the air, only spores. Spores are formed when a bacterium builds a wall around its genetic material for safekeeping in times of extreme dryness or the absence of food. Once the formation is complete, the rest of the cell disintegrates, releasing the spore. Spores may exist for decades before germinating into new cells. Bacterial spores form in a complex process either inside the parent cell or at its tips. In the micrograph, round bacterial spores can be seen lying dormant amidst fully active rods and filamentous forms (Figure 5.2).

Cysts are resistant structures, like spores, but made by protists. Many sorts of nongrowing propagules—roughly spherical, walled structures some five microns or larger—all have been dubbed "cysts," although it is likely that the many kinds of cysts so far studied are not directly related to each other but represent similar responses to threats of starvation and desiccation. For some reason, many protists don't make such protective structures unless they find a sexual partner first. In times of crisis, protist partners merge and membranes grow around the entire merged organism and preserve it. If humans build transparent domes around major cities to avoid environmental threats, we will in a sense be repeating developments that occurred in simpler organisms long ago.

Actually, spores and cysts are only two examples of survival strategies. They are paralleled in the plant kingdom by hard seeds and fruits. But each organism alive today represents the absolute success of all its ancestors that possessed certain survival strategies. Although modern organisms may show various degrees of divergence from ancestral forms, only extinct organisms, whether hominids or microorganisms, deserve to be considered less evolved. And even then the question is debatable because, according to the theories of the University of Chicago paleontologists Jack Sepkowski and David Raup, evolution may be a far more random process than formerly believed. If, as Sepkowski and Raup believe may be the case, the biota has been periodically

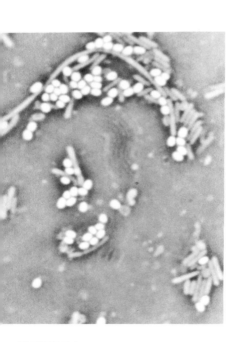

FIGURE 5.2

Bacillus bacterial spores. The spores are white and opaque; the growing bacterial cells are grey and rod-shaped.

decimated by meteorites that caused mass extinctions, then evolutionary survivors are not necessarily here by virtue of any special qualities. We simply exist today because our particular collections of genes, like stacks of chips growing on a roulette table, have not yet exhausted their winning streaks.

▪ SEX AND GENETIC RECOMBINATION ▪

What is sex? How did it evolve? By looking at the wide variety of sexual modes in the microcosm we can formulate ideas about the origin and importance of sex.

First we must define what we mean by sex and realize that sex takes on distinct forms in different organisms. Biologically, sex is simply shorthand for gene exchange—the deployment of DNA from the body of one to another organism. Sex is not related to reproduction except in certain cases. Because human beings, as well as most animals, represent one of these special cases, we tend to confuse sex with reproduction and not recognize them as separate processes.

Although some scientists think gene exchange predated the evolution of the first bacterial cell, bacterial sex probably evolved after cellular reproduction was in full swing. Because no ozone layer protected the Earth more than three billion years ago, ultraviolet radiation posed a dangerous threat to surface life, especially to photosynthetic life that needed the sunlight. Microbial mats, hardened into stromatolites and other layered community structures, were one way early life protected itself against uv light. Like ants crossing water on the backs of their drowning mates, some microbes sacrificed others for use as a living shield. But an even more successful response was the development of bacterial DNA repair mechanisms to fix uv-damaged DNA. In standard DNA repair, an organism copies an intact strand to replace a damaged part of the molecule. Such splitting and splicing is closely related to chromonemal sex—the merger of bacterial genetic material (Figure 5.3). Today, mutant bacteria that lose the ability to deal with uv light also often lose their entire capacity for sexual gene exchange.

Our genetic engineering technology is based on the strange sex of bacteria, which have been doing their own biotechnology for eons. Plasmids, viruses, and episomes are just some of the names that have been coined to describe the arsenal of DNA bits and pieces that are traded in bacterial sex. Bacteria are usually asexual reproducers; in the absence of danger their cells simply divide: one becomes two, two become four, four become eight, and so forth. In bacterial sex, however, a string of cell material connects two organisms and a variable amount of DNA flows from one cell—the donor— into the other—the recipient. The result is fewer genes in the one parent and more in the other. There is no offspring per se, just a recombined parent. Bacteria that trade away most or all of their genes die soon after the transaction. But the recombined survivors, having inherited new traits and abilities, may be the only organisms able to live in new environments. For example, imagine a bacterium vulnerable to attack by antibiotics but able to resist desiccation by producing spores, and another, sporeless bacterium that

FIGURE 5.3

Mating bacteria. DNA is presumably being passed from donor (left) to recipient (right). Electron micrograph.

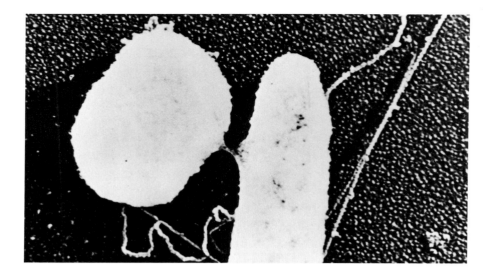

can grow in the presence of an antibiotic. If both kinds are exposed to an antibiotic-rich, rapidly drying environment they will all die except those that have transferred their genes to form new antibiotic-resistant, spore-forming cells. These recombinants will survive and reproduce.

This sort of gene transfer has been called "horizontal" and contrasts with the "vertical" sex of larger, sexually reproducing organisms. The great difference between the two types of sex is that larger organisms generally must produce offspring in order for genetic recombination to occur. Then, when offspring are produced, any improved combinations stand a good chance of being immediately diluted because, to reproduce such traits, the offspring must recombine yet again. Contrast this with horizontal sex and cloning: a bacterium receives genes via a partner, and then proceeds to clone (asexually reproduce) itself. The bacterium directly inherits and maintains possibly crucial new traits or metabolic abilities with impressive fidelity. This ability to adapt quickly to changing environments and to pass on that ability in an unmixed form is probably a major mechanism of planetary environmental control, and it lies at the heart of our infant industry of genetic engineering.

The remarkable maintenance of Earth as a chemically unstable system has been credited in part to horizontal sex. Bacteria probably have regulated the highly reactive gases of the atmosphere for millions of years at the least by virtue of their rapid growth and fluid gene exchange. While plants and animals certainly release and incorporate major amounts of oxygen and carbon dioxide, certain compounds such as methane, ammonia, hydrogen sulfide, and nitrogen can be transformed only by bacteria. These compounds are totally out of proportion to their expected values in a chemically stable system. The growing, gene-trading, burping, breathing bacteria are suspected to be living control mechanisms in a system of global elemental cycling and gas exchange.

From the point of view of planetary life, the vertical sex of people and other large living things is less glamorous. Although its praises have been shouted by humankind, meiotic sex is just an addendum to the functioning of the biosphere. Appearing only in the last billion years or so, meiotic eukaryotes showed up in a world already prepared for them by gene-exchanging prokaryotes.

74

As we have said, the meiotic sex of plants and animals is hitched to reproduction, it is different from the gene splicing of bacteria. In meiosis, genetically reduced cells result, containing exactly half the number of chromosomes as their parents. At some later moment these haploid cells will come together in fertilization to close the meiotic sexual cycle. In protoctists, this cell, sometimes a fertile egg, then divides by mitosis to produce a differentiated individual. Alternatively, in some protoctists and virtually all fungi, this fertile "egg nucleus" reverts to its haploid state by immediately undergoing meiotic cell division again. Sometimes meiosis occurs (or is assumed to occur) inside thick-walled cysts.

Many scientists have wondered why sexual organisms go through all the trouble of dividing their chromosomal numbers every generation only to double them again in fertilization. An asexual organism bypasses this process completely and can reproduce much faster as a result. Traditionally, the fact that sexually produced offspring supposedly can adapt more quickly to changing environments and have a greater variety has been used to explain their existence. Detailed evidence, however, has challenged both the notion that sexual reproducers show more variety than asexual organisms and the assumption that they adapt more quickly to unpredictable environments. The real reason why sexual organisms are still around remains a mystery. It may be because tissue differentiation depends on a subtle aspect of meiosis. While there are parthenogenetic animals (female animals that can produce fertile eggs that give rise to another female), no animal produces offspring without first going through the chromosome pairing aspect of meiosis and then forming egg cells.

The evolution of meiotic sex—which occurred separately in plants and animals—seems improbable. One thing is clear, however: meiosis developed from mitosis and both originated in the kingdom Protoctista. Protists exhibit the greatest number of variations not only of meiotic sex but of mitosis and nuclear arrangements (see Sea Whirlers, Chapter 11).

A plausible idea for the origin of meiosis suggests that eukaryotic cells ate each other when threatened with starvation, thereby becoming double. Their membranes and nuclei fused into a single cell with twice the ordinary number of chromosomes. This was almost fertilization. Then, when their chromosomes divided in the process of growth, each set did not replicate before moving to the offspring cells so that the formerly fused (double) cell was reduced to its original status. These earliest recoveries from the then abnormal diploid condition were a sort of meiosis. During times of desiccation, cold, or starvation, cycles of cannibalistic fusion followed by delays in intracellular timing that rid the monster cell of its doubleness would have led to the first forms of meiotic sex. Meiotic sex, then, may have served as an organizing tool in the development of highly complex, multicellular beings. These beings were destined to become doubled monsters when the sperm nucleus fertilized the egg nucleus, just as they were destined to be reduced to single cells in order to renew the perpetual cycle in each and every sexually reproducing generation. If you really want to know about the multiple origins and meanings of microbial sex, read our book listed on page 94.

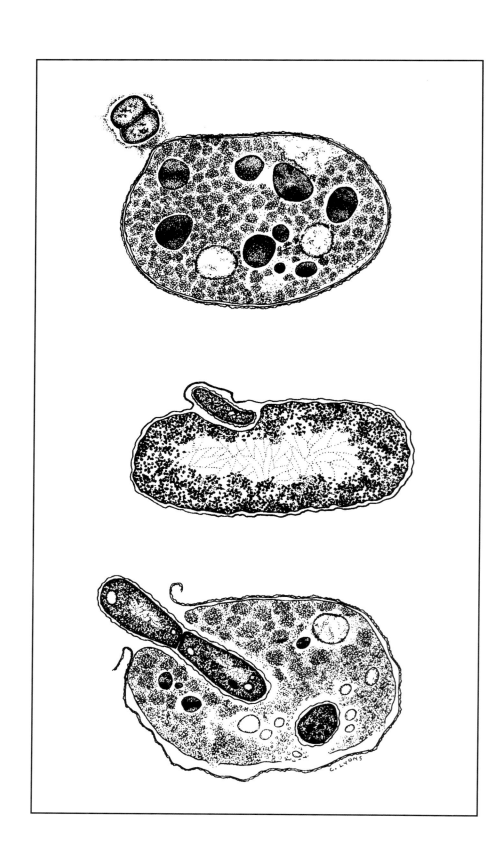

FIGURE 6.0

Bacterial predators (top to bottom): *Vampirococcus* attaching to *Chromatium*, *Bdellovibrio* entering a Gram-negative *E. coli*, and *Daptobacter* invading *Chromatium*. (Drawing by Christie Lyons.)

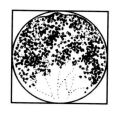

S·I·X

MICROBIAL INTERACTIONS

▪ PREDATION AND DEFENSE ▪

As in the human world of global politics, methods of attack and defense, arms races, and alliances have evolved in the microcosm. Bacteria with such colorful names as *Bdellovibrio* and *Vampirococcus* can attack, devour, and grow, killing an entire population of other bacteria, such as spirilli or pneumococci. Deleterious, even disastrous, relationships apparently can evolve into pacts and living partnerships. The illustration (Figure 6.0) shows *Vampirococcus* at the top and *Bdellovibrio* in the middle. At the bottom we see *Daptobacter* munching through the common purple bacterium *Chromatium*. *Daptobacter* bursts *Chromatium* apart as it divides inside the cytoplasm of the larger photosynthetic bacterium. But the deadliest parasites are actually the most ineffective: they wreck their own chances for survival with the host's body.

Bdellovibrio bacteriovorans are bacteria that must invade and reproduce inside other bacteria to survive. Taken from soil, lake, and sea, several different sorts of *Bdellovibrio* have been isolated into cultures. After exploiting (by digestion) all of the host material in order to multiply themselves, the *Bdellovibrio* bacteria burst out, spreading and entering new hosts in the local affected population. In part because of its tiny size relative to other bacteria, combined with the huge numbers of different bacterial hosts it is capable of invading, *Bdellovibrio bacteriovorans* has not yet run up against the inevitable problem of its destructive lifestyle, that is, killing off its last living habitat.

Oxygen-using bacteria like the deadly *B. bacteriovorans* probably were instrumental in the cellular invasion that led to the mitochondria of our cells. Free-living respirers occupied cells and became mitochondria—eternal fixtures within the cellular machinery of eukaryotic cells. They are now utterly dependent on the cells they once exploited. Similarly, microbiologist Kwang Jeon has observed the transformation of another bacterial parasite from deadly pathogen to symbiotic organism necessary to the survival of what can only be described as a new species of *Amoeba proteus*.

77

While many microbes are skillful predators, they also have developed numerous defense strategies. Like nations, bacteria have developed means of chemical warfare. Streptomycin and other antibiotics derived from bacteria are extremely effective in combating the growth of a wide spectrum of other bacteria and for this reason have been borrowed by man in the battle against disease. Many such antibiotics, including the famous penicillin, made by the ascofungus *Penicillium*, work by sabotaging the ability of bacteria to make new cell walls. At first people did not research and develop antibiotics, but only discovered their use against unwanted bacterial foes. Now, in our pharmaceutical prowess, we chemically modify the antibiotics to make them more soluble in our stomachs or more resistant to our stomach acids. Research has domesticated these microbes: they are grown in large vats and encouraged to produce the antibiotic starting materials for us as they have for themselves, very effectively, for millennia.

Although most metabolic pathways, and thus chemical defenses, have their roots in bacteria, the structurally more complex eukaryotic cells have devised quite intricate mechanical attack and defense strategies. Some protists form poisoned darts called trichocysts or toxicysts with which they stab their prey (Figure 6.1). Among the members of the kingdom Protoctista are agents of some of the most deadly and debilitating diseases known to man, woman, or beast. Many of these, such as malaria and trypanosomiasis (sleeping sickness), are tropical diseases. The complex life cycles of such parasites often involve inhabiting a series of hosts belonging to different species, making such parasites particularly difficult to follow and defend against.

But the protoctista pioneered beneficial as well as deleterious associations. From an evolutionary standpoint, one or two lines of protoctists evolved into the first animals, which can be seen as intensely coordinated consortia of bacteria. From this perspective, the development of calcium carbonate shells and of calcium phosphate bones can be seen as another page in the history of microbial attack and defense. Calcium is always used in the electrical metabolism of cells with nuclei. A buildup of calcium on the inside of the cell, however, causes death and so the first eukaryotic cells to survive were forced to release excess calcium through their membranes into the surrounding waters. But, as protists came together to form multicellular protoctists, and as the first animals evolved from these protoctists, relatively large quantities of hard calcium compounds got deposited outside. Cell collectives, extruding calcium carbonate in ordered arrays, produced a kind of "armor." Long before medieval knights and horses crossed Europe in metal armor, eukaryotic cells built castles and shields in the form of animal exoskeletons. Fossilized hard parts demonstrate the durability of these solid structures constructed by clones of nucleated cells.

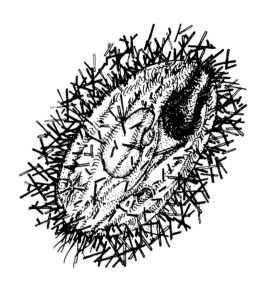

FIGURE 6.1

"Poisoned darts" or trichocysts (a kind of extrusome) that have been extruded by the microbial mat ciliate *Tubigula kahli*. (Drawing by J. Steven Alexander, based on studies by Betsey Dyer.)

▪ ACQUISITIONS AND MERGERS ▪

One of the major ways by which different organisms maintain themselves and organize into close-knit communities is by feeding. Assemblages occur as some species feed and prosper on the refuse of others. Many animals, such as hydras, marine snails, flatworms, and giant clams, eat algae "vegetables" that are not digested but continue to photosynthesize right through the translucent animal tissue after being eaten

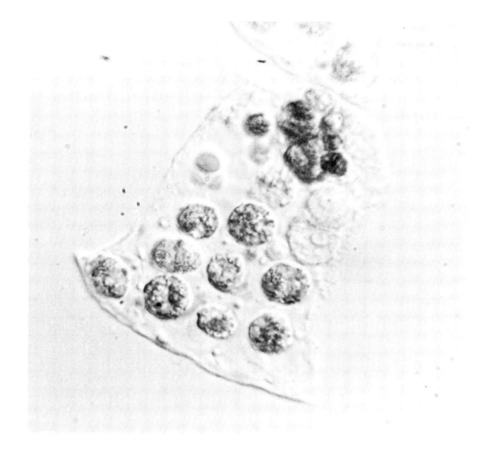

FIGURE 6.2
Spherical dark dinomastigote photosynthesizers living within a translucent surface tissue cell of a coral. (Photo courtesy of R. Trench.)

(Figure 6.2). The photosynthetic activity provides a surplus that feeds the animal from inside. In the case of the green flatworm *Convoluta roscoffensis,* uric acid waste made by the worm is completely recycled as useful amino acid nitrogen for the algae.

The same sort of relationship occurs in an even more dramatic fashion in the microcosm. Organic acids are expelled by such photosynthetic organisms as the algae of lichens and used by the fungi of lichens until nutritional collectives are formed—the adult lichen itself—that make almost complete use of available chemical compounds. Coral reefs, in which all coral animals harbor symbiotic dinomastigotes, provide a spectacular example of an immensely productive ecosystem thriving in the midst of the nutrient-depleted waters of the world's tropical oceans.

Motility associations also are prevalent in the microcosm. As we have seen in the hindguts of termites, for example, swarms of small spirochetes can be seen pushing protists many times their size. Some spirochetes attach to and feed off the surface of other symbionts. In some cases, such as that of the free-living protists *Mixotricha paradoxa* and some species of *Pyrsonympha,* individual spirochetes attached and feeding at the surface cannot be distinguished from undulipodia without the aid of an electron microscope.

From time immemorial, the human myth-making imagination has been enchanted by chimerical creature combinations: the sphynxes of Egypt; the flying, fire-breathing dragons of China; the winged horse, Pegasus. From animal recombinations to the angels and devils of Christianity, to vampires, fairies, Apollo, and Icarus, people have been intrigued by the idea of borrowing from birds (or bats or dragonflies) in a skyward expansion.

Spirochetes may have in reality accomplished a revolution in transportation when they joined up with slower, clumsier bacterial consortia and gave their companions increased powers of movement. Territorial expansion led to further opportunities for development. Covered with spirochetes capable of penetrating the most viscous fluids, these composite beings opened up new environmental niches.

Whether or not one believes the spirochete hypothesis, it is sure that feeding and motility associations, both loose and strict, occur now and have in the past. In the watery, multibillion-year-old environment of the microcosm, such associations have brought cells inside of each other. Sheltered and fed inside other beings, organisms have flourished—more so than they would have had they been left to their own resources.

We return to the termite hindgut because it is a visible example of the complexity of biological relationships at a variety of levels. Within the supposedly sovereign individual termite, itself feeding on wood, thrives a bustling throng of fantastic microbes that digest the wood; they form motility alliances and nutritional partnerships. The first description of the termite microcosm still retains a certain freshness today. The inquisitive 19th century American physician Joseph Leidy was elated by his discovery of the complex intimacy inside the overgrown hindgut of the termite *Reticultitermes flavipes*. In his publication *The Parasites of the Termite*, Dr. Leidy described the wonderland revealed by his microscope:

> *In watching the Termites from time to time wandering along their passages beneath stones, I have often wondered as to what might be the exact nature of their food in these situations. Observing some brownish matter within the translucent abdomen of the insects, I was led to examine it with the object of ascertaining its character. On removing the intestinal canal of an individual I observed the brownish matter was contained within the small intestine, which is comparatively large and capacious. The brownish matter proved to be the semi-liquid food; but my astonishment was great to find it swarming with myriads of parasites, which indeed actually predominated over the real food in quantity. Repeated examination showed that all individuals harbored the same world of parasites wonderful in number, variety and form.*

Dr. Leidy's discovery of the crowded microbial community that actually digests the tough cellulose of the wood eaten by termites can be repeated by anyone with any sort of microscope. The wondrous details of trichonymphids, prysonymphids, and dienymphids can be resolved with a compound light microscope of even 100 magnification power. As Leidy put it,

> *If the intestine is ruptured, myriads of the living occupants escape, minding one of the turning out of a multitude of persons from the door of a crowded meeting-house. So numerous are the parasites and so varied their form, movement, and activity, that their distinctive characters cannot be seen until they become more or less widely diffused and separated.*

With a bit of your own saliva you can make a glass slide and coverslip preparation of microbial inhabitants that will last for hours. If you can find some wood-eating termites you can observe the peculiar protists that are limited to insect intestines. Other insects may contain other curious symbionts as well. Indeed, it was Anthony van Leeuwenhoek's discovery of tiny creatures parasitic on fleas that inspired the great Anglo-Irish satirist Jonathan Swift to write:

> So, Nat'ralists observe, a Flea Hath smaller Fleas that on him prey, And these have smaller Fleas to bite 'em, And so proceed ad infinitum.

Perhaps Swift found in the work of Leeuwenhoek and others inspiration for *Gulliver's Travels*, written to "vex the world," and which contained the land of Lilliput whose civilized inhabitants were so tiny they could fit in the palm of Gulliver's hand.

USING · THE · FIELD · GUIDE

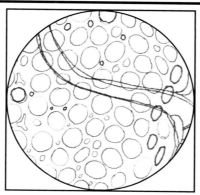

FIGURE 7.0

All of these organisms form stalks, yet each belongs to a different kingdom. Shown are *Hydra*, an animal; *Polytrichum*, a plant; *Amanita*, a fungus; *Chrondromyces*, a colonial bacterium; and *Sorogena*, a ciliate protoctist. (Drawing by J. Steven Alexander.)

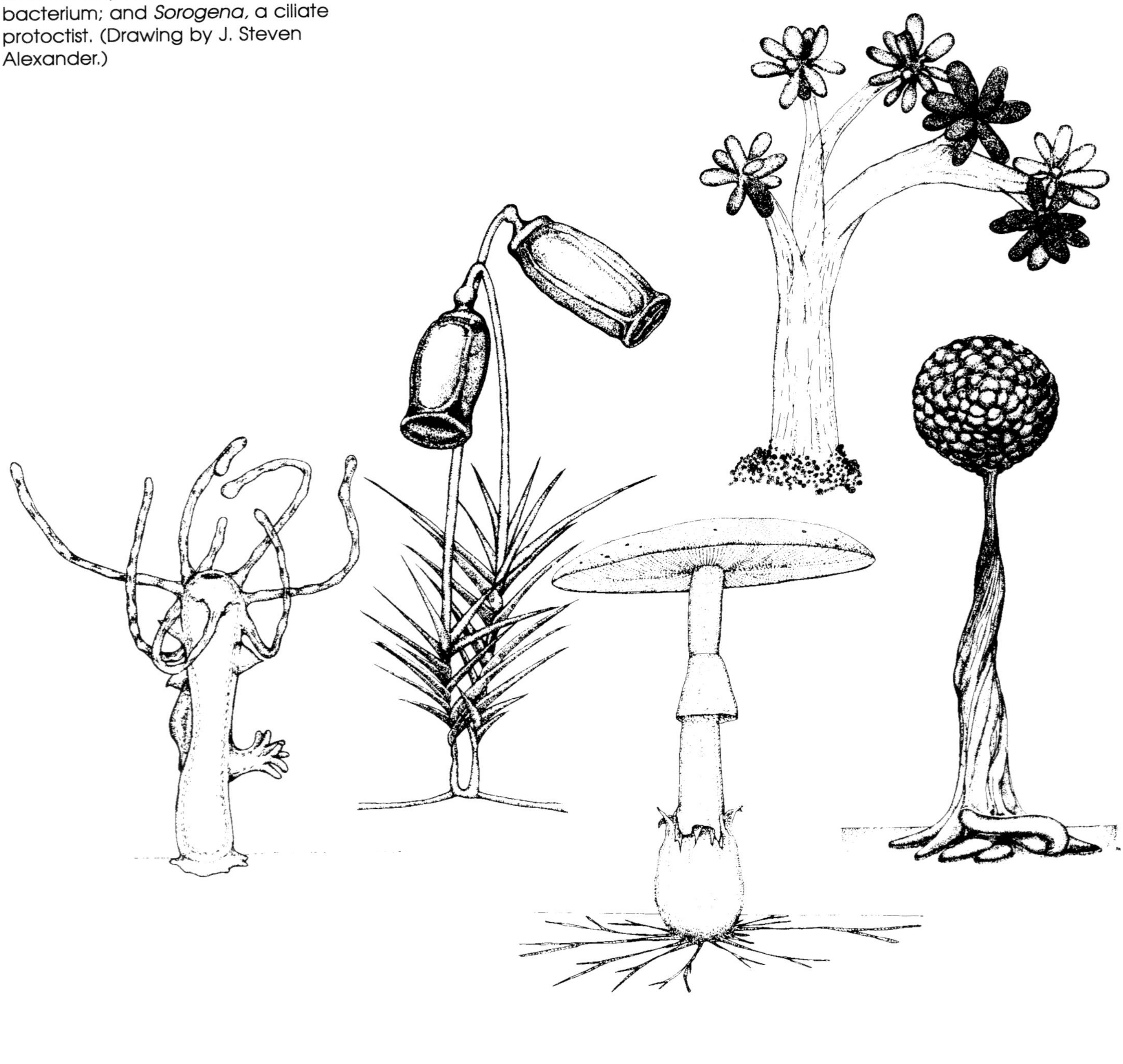

NAMES

▪ INTRODUCTION TO THE NOMENCLATURE ▪

One purpose of this book is to provide a new framework within which we can view the various life forms of the microcosm. Usually, such a framework includes naming organisms and identifying their attributes and patterns of similarity. But naming organisms sets limits both on how we see them and on the scope of what we can see and say about them. Names are artificial, imposed by man to organize and understand the world around him. In reality, nature moves without respect for the limits imposed by human definitions. Nonetheless, we need names to talk about organisms, groups of organisms, and their ecological connections.

The microscope gave a dark world windows through which scientists could examine the large and the small: animals, plants, algal forms, fungi, protozoa, even "germs." Through the microscope it becomes apparent that all organisms evolved from the same ancestors very early in the history of life on Earth. And yet, while cellular structures and microorganisms often are identical, they frequently have different names in different disciplines and the names are often inaccurate. For example, in "Protozoology," the Greek root *zoos* means animal, suggesting that colorless microscopic beings are primitive animals even though not one of them develops from the fusion of egg and sperm. Euglenids, sea whirlers, and all sorts of members of the microcosm have been co-opted by "protozoologists" while these same organisms have been placed in entirely different "families" by botanists. It seems better to call these free-living, nucleated cells protists, and to include them with their descendants in a kingdom of their own, the Protoctista. In fact, those who study such beings have begun to call themselves protistologists or protoctistologists, rather than protozoologists.

Numerous inappropriate names flourish: "flagellates" is another example. Since only bacteria have flagella, it is not consistent to call eukaryotes with undulating organelles "flagellates." Instead we should call them mastigotes, an old trusty name for cells with nuclei and that move by undulipodia. Dinoflagellates, for example, should then be referred to as dinomastigotes.

The field of nomenclature is not easily changed. Although they are less confusing, new names are not so widely known as their older counterparts, and in order to look up information on them it is often necessary to know the old names as well. We have taken the liberty in this guide of replacing long-winded scientific names with shorter English equivalents. But when new names are used for the first time we give the old names in parentheses.

Protistology as a combination of protozoology, phycology, mycology, and other disciplines is just coming into focus in its own right. [A meeting of the very young International Society for Evolutionary Protistology (ISEP) was held in 1987 in Ascot, England. This society continues to meet in Europe or North America every two years.] Nomenclature among new protistologists is still hotly debated. Classification and naming of unfamiliar microbes is in a state of flux. We have tried to use names that are consistent with both the five-kingdom view and the experience of amateurs and professionals who tour the garden of microbial delights.

▪ FIVE KINGDOMS ▪

Monera, Protoctists, Fungi, Animals, and Plants

According to the five-kingdom classification scheme, the kingdom Monera (bacteria) is the first. Monera consists entirely of prokaryotes, or organisms composed of cells without nuclei. Some authors prefer to call this the kingdom Prokaryotae. We prefer Monera, however, because we use Prokaryotae as an appellation for the superkingdom of bacterial forms that stands in opposition to the Eukaryotae and its four kingdoms whose members are composed of cells with nuclei (Figure 7.0).

The second kingdom, Protoctista, contains organisms that first appeared as fossils at least 1500 million years after the appearance of bacteria. Protoctista is a catch-all group of miscellaneous creatures and the least well-defined kingdom of the five. Protoctists include all nucleated microorganisms that are not plant, animal, or fungi. Many members are multicellular: individuals can be composed of many cells even though protoctists lack embryos and none show the tissue level of organization. Protists are single-celled protoctists.

Further study of this kindgom is likely to reveal clues to the origin of cell organelles, multicellularity, cell motion, mitosis, two-parent meiotic sex, and life cycles. The protoctists are capable of a wide range of activities: we find built-in "weapons," strategies of predation, desiccation resistance, photosynthesis, cell differentiation, colony formation, transportation networks, "house" construction, food processing, and "walking." The field guide that follows will depict much of this diversity.

Members of the third kingdom, Fungi, are eukaryotes that grow from spores, never form embryos, and are amastigote (lack undulipodia) at all stages of their life cycle. Fungi have morphogenetic and intracellular motility, but they never move around by themselves. Fungi are osmotrophic organisms: they do not photosynthesize or ingest food through a mouth but absorb nutrients from water or from the tissue of protoctists, animals, or plants.

Animals, of course, represent the best-known kingdom. In two-kingdom classifications, animals are simply creatures that move. Metazoa are large, multicellular animals, whereas protozoa are the so-called one-celled animals.

86

But in the five-kingdom classification system animals are more rigorously defined. They are multicellular organisms with two sets of chromosomes in each cell that develop from the fertilization of a large egg by a small sperm. The product of fertilization of egg by sperm is a zygote; a fertilized egg that grows by a series of mitotic divisions into an embryo is called a blastula. In the five-kingdom classification this hollow ball made of many cells, the blastula, is common to all animals.

Like those of the kingdom Animalia, members of the kingdom Plantae are multicellular. Plants are sexually reproducing eukaryotes that develop from an embryo supported by the mother's nongrowing tissue. Unlike an animal embryo, however, plant embryos are never blastular. In the most familiar cases the plant embryo develops into a little bump inside a seed. In addition, cells of most plants contain green chloroplasts with two kinds of chlorophyll (chlorophyll *a* and *b*) as well as yellow and red pigments, such as xanthophylls, all of which are involved in the photosynthetic process. Although often green, photosynthetic, and multicellular, eukaryotic protoctists are never plants because, unlike true plants, they never grow from embryos.

■ TAXONOMY ■

Carolus Linnaeus (1707–1778) erected the useful system of double Latin species names (such as *Homo sapiens*, humans, and *Agaricus campestris*, a field mushroom) called binomial nomenclature. Linnaeus instituted his highly successful system to provide an orderly and international scientific vocabulary to refer to organisms that before his time had many vernacular or common names. The first part of a Linnaean title refers to the genus. For instance, people are in the genus *Homo*, meaning man. Although fossil hominids exist, no other live organisms are in our genus. The second name in a Linnaean title refers to the species—defined in animals and plants as a group of organisms whose members are morphologically very similar to each other and can only breed with each other. In *Homo sapiens* the species name is *sapiens*. Like the English words sage and sapient, sapiens comes from the Latin *sapere*, meaning to taste, to have good taste, or, more to the point, to be wise.

Systematists, however, have augmented Linnaeus's original system to include a whole hierarchy of taxonomic levels above genus and species. From the top down we have kingdom, phylum, class, order, family, genus, and species. People are animals in the kingdom Animalia, phylum Chordata, class Mammalia, order Primates, family Hominidae, genus *Homo*, species *sapiens*. New or intermediate taxonomic levels are sometimes formed by adding "super" or "sub" as prefixes to the name of a given taxonomic level. For instance humans also are members of the subphylum Vertebrata, as well as the superkingdom Eukaryotae. If some human beings settled Mars and, after living there for many generations some returned to find they were so different they could no longer sexually reproduce with people on Earth, Martians might be considered a new human species. Because all peoples—*Homo sapiens sapiens*—alive on Earth today are ultimately accessible to each other, all coming from a common stock about 40,000 years ago, there are no human subspecies.

To recognize and classify organisms taken from nature is often easy. But some organisms are easier to handle than others. Tiny animals, such as brine shrimp and nematodes, are easier to recognize than protoctists, such as paramecia. Bacteria, because they are considerably smaller still, are far more difficult to classify.

The first step in identifying an organism is simply to get a good idea of what it is. Brine shrimp, for example, are "jointed foot animals," while nematodes are "round worms." With a picture, you may be able to identify your tiny shrimp as *Artemia*. At this point you have classified your organism at the genus level. With nematodes, however, you will not be so lucky—there are over 50,000 species described and only professionals can tell whether a sample has ever been classified. Positive identification would take detailed anatomical work and, if you can accomplish this, you will have become as skillful as a professional biologist. For ciliate identification, good observations coupled with picture-matching in books like *How to Know the Protozoa* may bring you to genus level. To go further and identify any but the largest ciliate species you probably would have to fix and stain preparations.

Clues to the members of the microcosm you actually see in your sample come from (1) geographical location and habitat where the organism was found, (2) its morphology, in other words, the shape, form, swimming behavior, etc., of the organism, (3) its physiology—for example, does it need oxygen, light, elemental sulfur? It helps to learn common names of organisms from friends and teachers. From this sort of informal information you must move to professional literature to find the Latin genus and species of your microscopic "pets." First it is necessary to determine if your sample has a Latin name and, if so, in which traditional literature it is registered: bacteriological, mycological, phycological, or protozoological.

Even professional biologists are frustrated by the paucity of field guides to the microcosm. The huge number of undescribed microbes and the difficulties of distinguishing between types of monerans and small protists demand of their classifiers painstaking work, including the gathering of physiological data, observations of growth patterns and life cycle modes, electron microscopic ultrastructure data, and even genetics. This sort of work on obscure microbes is severely hampered by the contradictory naming practices, the limited quantity and value of much of the literature, and the scarcity of expertise. In some subchapters of this guide we have had to leave out information about where and how to identify a particular genus because of a lack of data or because to do so would require sophisticated techniques beyond the range of our readers.

But there is a heartening side to these taxonomic gaps. Your chances of finding an organism *new* to science, especially in tropical muds and waters, are very good!

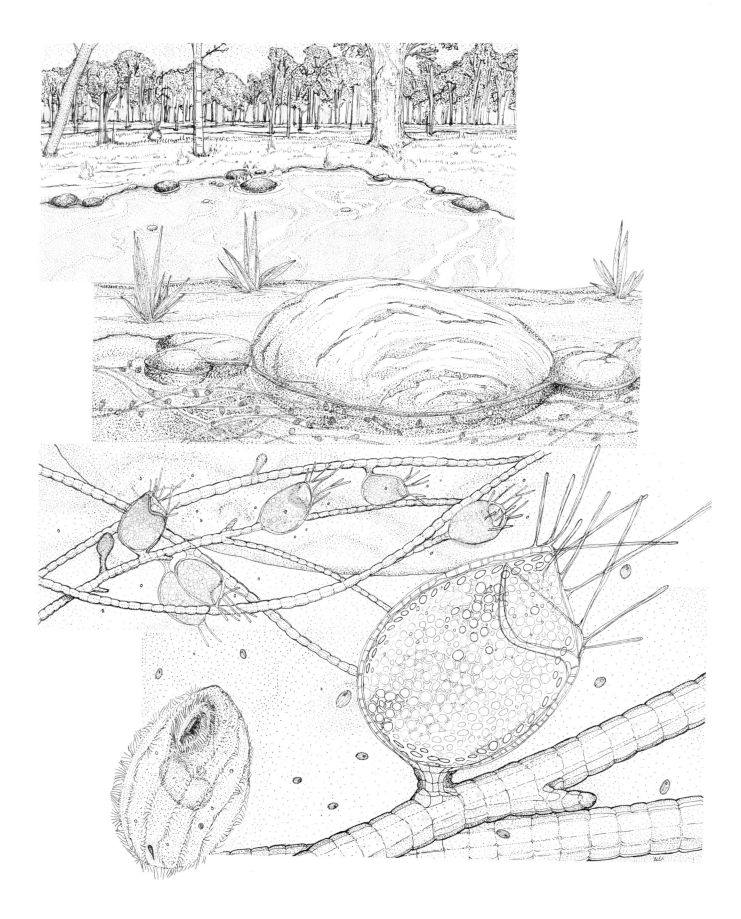

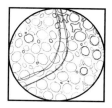

E·I·G·H·T

COLLECTING
AND
KEEPING YOUR MICROBE

▪ COMMUNITY LIVING ▪

icrobes make strange pets. Unlike dogs, cats, fish, or birds, it is
impossible to keep a single protoctist or bacterium. Microbes are
community organisms with more intricate biological connections
to other organisms than animals or plants. Thriving microbes
create environments that are attractive to the reproduction of other microbes.
Only in rare cases, and then with the specialized and complex techniques
of microbiology, can purebred populations or clones, starting from a
single cell, be grown.

But even the concept of pure cultures—of growing only one sort of
microbe by itself—is an idealization. This concept came about because
bacteria are so small, weighing on the order of a millionth of a millionth of a
gram, that it is possible to study the chemical processes they perform only by
measuring the effects of large numbers of similar ones. Yet new evidence
shows that even bacteria, reproducing only by cell division, show differences
from parent to offspring in every generation. The idea of "pure cultures"
permits microbiologists the idealized concept that the metabolism of billions
of interliving bacteria is the sum of the activity of individuals in such
collectives. Because of our need to study millions at once in order to see their
effects, only the metabolism of huge populations of bacteria is well known.
We understand individual microbes about as well as a Venusian observing
people in city traffic through a telescope would know about us.

To learn to grow only one sort of microbial type—that is, to make and
keep pure cultures of microbes—you'll need help from a book or lab manual
of general microbiology or bacteriology, such as those listed in the references
here. Many people, from college biology majors to nurses, biology tech-
nicians, and high school teachers, are familiar with this very useful "sterile
technique." Discuss your plans for keeping cultures of microbes with these
people; most of them will be glad to help you.

FIGURE 8.0
Utricularia (common name,
bladderwort), a plant found in
ponds and streams, traps microorganisms such as the ciliate *Tetrahymena* (lower left) in bulbous
bladders (lower right). The bladderwort then digests the ciliates and
uses their dead bodies as a
nitrogen source. (Drawing by J.
Steven Alexander.)

91

While the armchair naturalist may be perfectly content to read this book from the comfort of his or her home, readers who wish to explore the microcosm firsthand will need a compound microscope. Available on loan from well-equipped elementary and secondary schools, especially during the summer, compound microscopes also may be obtained at reasonable prices from dealers and scientific and educational supply houses. It is *always* preferable to buy a secondhand or new student microscope than a toy. Guides for the use of your microscope are readily available; if in doubt call a local biology teacher. If possible, also try to obtain access to a binocular "dissecting" microscope that magnifies from 20 to 50 times so that you can have a clear idea of what you are looking at before you probe more deeply into its fine structure. Failing this, get a good hand lens. In all the practical suggestions we make in this book we assume you have access to, and a little experience with, a microscope. But of course, if you prefer just to sit back in your armchair and relax, enjoying our tour of the manifold organisms of Earth, that is fine too.

▪ CARE AND FEEDING ▪

As medical and hygienic measures show, one often expends more effort to prevent the growth of microbes than to promote it. Nonetheless, particular microhabitats favor the growth of particular microbes. In microbiology, a medium is a surface or layer of substance in which a microbe lives and from which it derives nutrition. Traditionally, microbes have been grown in small circular plastic dishes called petri plates that contain a bottom layer of agar—an inert, organic hardened jelly to which such nutrients as sugars and salts are usually added. With appropriate nutrients, agar can support a wide variety of microbial populations and communities.

Agar is derived from a limited number of commercially important red seaweeds. Because it is smooth, hard, and colorless, it permits the observer to see clearly the growth of protoctists or colonies of bacteria. In order to culture microbes on agar a food source must be added because almost no bacteria, protoctists, or fungi can break down and use the agar as food. A single bouillon cube or eight-ounce can of V-8 juice added to a quart of agar liquid medium is fine as a non-specific food for microbes. Approximately 30 to 50 petri plates can be poured from such a mixture. Wheat seeds embedded in agar to which no other nutrient is added support good growth of fungi, such as *Coprinus*. Experimentation is needed, but agar media are useful and fun—the work is well worth the effort.

Of course, virtually any organic medium can support the growth of myriads of microbes. Try slightly dampening various articles in the bottom of a small jar. Cover the jar and leave the material for at least a few days. Pieces of fruit, joint compound, seaweed—even pepper or coffee grounds—tend to promote the rapid growth of certain kinds of microbial communities. Of course, moist soil and drops of pond water or oceanic detritus already have burgeoning microbial communities that can be discerned and directly examined with the use of a low-power light microscope. Cleaned peanut butter jars filled with pond water and placed on sunny window sills develop wonderful blooms of microcosmic inhabitants. If you are interested in enriching—capturing and growing—a specific sort of microbe, look up the microbe of interest to you in the Guide. We provide one example here (Figure 8.0): the search for the ciliate *Tetrahymena*, which lives and often dies inside the carnivorous trap of the bladderwort plant.

On the east coast of the United States, especially from the mid-Atlantic states to Florida, the bladderwort (*Utricularia*) lives at the edges of freshwater ponds and streams. Animals generally feed on plants but bladderwort, one of the fewer than 0.2% of plants that reverses this usual trend, feeds on animals. The bladderwort digests small animals and protists, including insect larvae, aquatic worms, water fleas, and ciliates, in order to compensate for chronic shortages of soil nitrogen. With the naked eye, the bladders look like a mess of dark tiny seeds; upon a closer look they appear to be miniature bulbs. The bulbous bladders, actually modified leaves that bulge out in pairs along the stem, act on the principle of lobster traps: what gets in can't get out. The actual mechanism, however, is more sophisticated. Through a physiological process, the water tension or pressure of the water inside the hollow bladders is less than that of the surrounding water. When a small swimmer approaches the closed valve at the entrance of the bladder and touches tiny bristles there, the valve opens: this draws in water and, with it, the animal or protist source of nitrogen. One of the most common ciliates caught by nitrogen-needing bladderworts is *Tetrahymena* (shown in Figure 8.0, blown up at bottom left), which feeds on the bacteria and plant juices leaked into the water inside the closed bulb of its captor. Because *Tetrahymena* are only slowly digested, you'll have plenty of time to watch them.

You need only your microscope, a medicine dropper, and a snipped-off bladder. To observe the bladder trap, just snip it for observation under low power. Alternatively, if you collect the plant, you can keep it alive for months in its original water in a terrarium or an aquarium that provides a bit of high ground or rock. Extract the juices from the snipped-off bladder with a medicine dropper, place the liquid on a slide, and at magnifications of 100 to 400 examine the world of trapped ciliates. You should be fascinated for days.

▪ HELP FROM THE LITERATURE ▪

We include here an annotated list of references to books that can help you isolate and grow certain kinds of microbes.

Aaronson, S. (1970). *Experimental microbial ecology.* Academic Press, New York and London. 236 pp. General microbiology text with procedures for growth, enrichment, and isolation.

Atlas, R.M. and Bartha, R. (1981). *Microbial ecology: Fundamentals and application.* Addison-Wesley Publishing Co., Reading, Mass. A college text responding to the need for general awareness of the pervasive microbial world. Contains many examples of microbe interaction with animals and plants, the practical uses of microbes, and references to useful general literature.

Behringer, Marjorie, P. (1973). *Techniques and Material in Biology*. McGraw-Hill Book Company, New York. A guide, especially for biology teachers.

Brock, Thomas D., Smith, D.W. and Madigan, M.T. (1984). *Biology of Microorganisms*. 4th ed. Prentice Hall, Inc., Englewood Cliffs, N.J. Entirely comprehensive college text book covering all aspects of microbiology. If you want to own a single encyclopedic reference, this is it.

Dixon, Bernard. (1976). *Magnificent Microbes*. Atheneum, New York. 251 pp. A fascinating account for the layman of many of the more important, beneficial roles of microbes.

Eklund, C. and Lankford, C. (1967). *Laboratory Manual for General Microbiology*. Prentice-Hall, Inc., Englewood Cliffs, N.J. 299 pp. A very basic, classroom text, with clear instructions and drawings for adult students taking microbiology for the first time.

Gerhardt, P., et al., eds. (1981). *Manual of Methods for General Bacteriology*. American Society for Microbiology, Washington, D.C. 524 pp. A comprehensive source. Includes sections on both light and electron microscopy (preparation, etc.), growth, taxonomy, and laboratory technique. Procedures for growing microbes in the absence of oxygen (anaerobes) are given.

Gest, H. (1987). *The World of Microbes*. Sci Tech Publishers, Inc., Madison, Wisconsin. 250 pp. A readable, informed first course in microbiology that explains standard techniques, the history of the science and, in general, provides a human context for "germs."

Gillies, R.R. and Dodds, T.C. (1968). *Bacteriology Illustrated*. 2nd ed. Williams and Wilkins Co., Baltimore. 198 pp. Nice drawings and photographs, easy-to-read text but primarily discusses bacteria of medical interest.

Grave, E.V. (1984). *Discover the Invisible: A Naturalist's Guide to Using the Microscope*. Prentice-Hall, Inc., Englewood Cliffs, N.J. 195 pp. A guide to the microbial world emphasizing microscopes—their history, their powers of revelation, and how they work.

Hale, Mason E. (1979). *How to Know the Lichens*. Win. C. Brown Co., Dubuque, Iowa. 272 pp. Well-illustrated, up-to-date, spiral-bound introduction to all the common North American lichens, showing their ranges and describing their appearance.

Jahn, T.L. (1949). *How to Know the Protozoa*. Dubuque, Iowa: Wm. C. Brown Co. 234 pp. An old, trusty, spiral-bound introduction to the nonphotosynthetic protoctists with excellent, simple drawings and explanations. Fine for genus and species, but written using the framework of the old animal–plant dichotomy in which all the organisms discussed are considered animals.

Margulis, L. and Sagan, D. (1991). *Origins of Sex: Three Billion Years of Genetic Recombination*. Yale University Press, New Haven CT. 258 pp. Illustrated account of the evolution of bacterial recombination and meiotic sexuality in protoctists. Haplo-diploid sex, differing entirely from molecular replication and organism reproduction, evolved several times at least in our sexy protoctist ancestors. For teachers and serious students.

Margulis, L. and Schwartz, K. (1988). *Five Kingdoms: An Illustrated Guide to the Phyla of Life on Earth*. 2nd edition W.H. Freeman and Company, New

York. 376 pp. Comprehensive guide to phyla of bacteria, protoctista, fungi, animals, and plants.

Olive, L.S. (1975). *The Mycetozoans.* Academic Press, New York, San Francisco, and London. The definitive work on the biology of the slime molds.

Prescott, G.W. (1964). *How to Know the Fresh-water Algae.* Win. C. Brown Co., Dubuque, Iowa. 272 pp. Excellent, simple drawings. Easy to use but, as Jahn's book, limited to the plant–animal conception.

Primrose, S.B. and Wardaw, A.C. (1982). *Sourcebook of Experiments for the Teaching of Microbiology.* Academic Press, New York. Comprehensive and useful, especially for bacteriology fans and teachers.

Seeley, H.W. and VanDemark, P.J. (1972). *Microbes in Action: A Laboratory Manual of Microbiology.* W.H. Freeman and Co., San Francisco. 361 pp. Basic classroom text with clear drawings and descriptions of the classical demonstrations and experiments performed by microbiology students.

Sieburth, John McNeill. (1979). *Sea Microbes.* Oxford University Press, New York. Written by a professional microbial ecologist, this beautifully illustrated, coffee table guide displays the richness of the marine microcosm. Especially striking are the microbial seascapes photographed using the scanning electron microscope.

Sonea, S. and Panisset, M. (1983). *A New Bacteriology.* Jones and Bartlett Publishers, Inc., Boston, Mass., and Portola Valley, Calif. 140 pp. The best introductory statement of the ecological and evolutionary role of bacteria without the lapses into highly technical material typical of microbiology textbooks.

▪ BIOLOGICAL SUPPLY HOUSES ▪

If you know what you want but can't get it, the following three major biological supply houses are likely to have it available. We include their addresses for the sake of completeness.

Carolina Biological, 2700 York Road, Burlington, North Carolina, 27215.

Tuttox, 8200 South Hoyne Avenue, Chicago, Illinois, 60620.

Ward's Natural Science Establishment Inc., P.O. Box 92912, Rochester, New York, 14692-9012, 800-962-2660. FAX 800-635-8439.

In addition, the American Type Culture Collection Catalog of Strains, published by American Type Culture Collection, 12301 Parklawn Drive, Rockville, Maryland, 20852-1776, has all listings of media and microbes from bacteria to fungi and protists. The organisms offered in this catalog are for sale live. The University of Texas Culture Collection in Austin, Texas, can provide you with samples of live algae.

Ward's also supplies our 35mm color slides of bacteria, protoctista and fungi in "The Five Kingdom Slide Set." The Five Kingdom poster and teaching activites relevant to growing your microbes are also available.

For a complete listing of living and fossil protoctists and guide to the professional literature see:

Margulis, L., Corliss, J.O., Melkonian, M., and Chapman D.J., editors (1990). *Handbook of Protoctista. The structure, cultivation, habitats, and life histories of the eukaryotic microorganisms and their descendants exclusive of animals, plants and fungi. A guide to the algae, ciliates, foraminifera, sporozoa, water molds, slime molds, and other protoctists.* Jones and Bartlett Publishers, Boston MA.

Margulis, L., McKhann, H.I., and Olendzenski, L., editors (1993). *Illustrated Glossary of Protoctista: Vocabulary of the algae, apicomplexa, ciliates, foraminifera, microspora, water molds, slime molds, and the other protoctists.* Jones and Bartlett Publishers, Boston, MA.

Illustrated Glossary of Protoctista contains comprehensive lists of sources of live material available worldwide. Names and addresses of research scientists, experts on live and fossil protoctists, are also listed.

THE · GUIDE · TO · THE · MICROCOSM

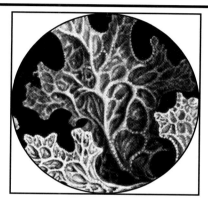

INTRODUCTION
TO THE GUIDE

H aving briskly toured the vast microbial garden and its history, we are ready to return to make a more leisurely study of its many inhabitants. We divide all microbial life into the three kingdoms of bacteria, protoctists, and fungi. These appear as Chapter 10 (Bacteria), Chapter 11 (Protoctists), and Chapter 12 (Fungi). Our order is roughly chronological and we include the most commonly encountered creatures in our descriptions.

Although the first bacteria were present on Earth approximately 3500 million years ago, the first protoctist probably did not appear until well after 2000 million years ago. Because the evolution of fungi is so connected with the appearance and diversification of plants, it is doubtful that fungi appeared much before 450 million years ago. By this date, some desiccation-resistant green algae had already evolved into the ancestors of all land plants. By now there are estimated to be some 20,000 different kinds (biotypes) of bacteria, 200,000 species of protoctists, and 100,000 fungi species.

In recognition of the dynamic nature of any classification system, we present here current views of subvisible organisms. Our guide does not include either tiny animals, such as rotifers or nematodes, or tiny plants, such as duckweed, even though these organisms live at the border between the microcosm and the visible macrocosm. (Some protists, such as ciliates, feed on tiny rotifer animals.) Many bacteria, protist, and fungi types are not included here in our short descriptions. They remain elusive because of their mysterious or obscure nature. They await discovery, perhaps by you.

FIGURE 9.0

Various types of viruses. Two types of cold viruses, other human viruses, and a DNA virus of bacteria. (Drawing by J. Steven Alexander.)

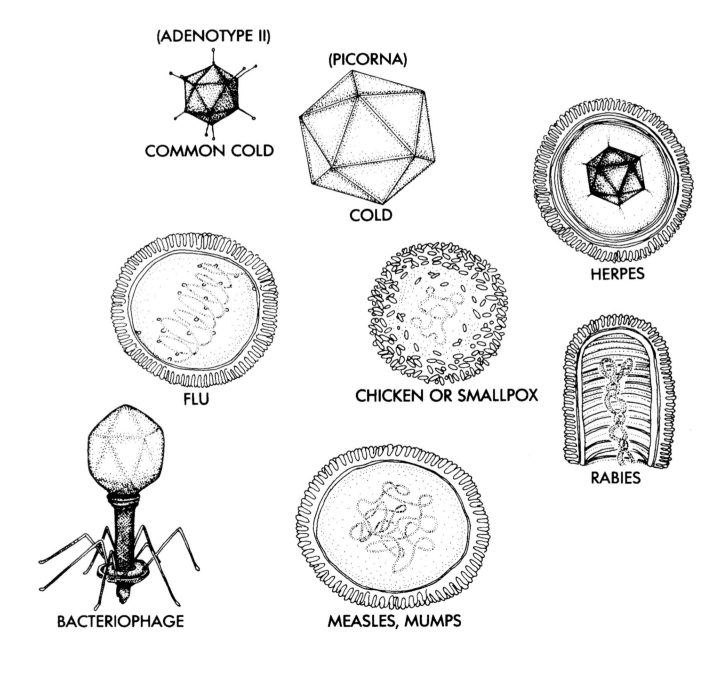

(ADENOTYPE II)

COMMON COLD

(PICORNA)

COLD

HERPES

FLU

CHICKEN OR SMALLPOX

RABIES

BACTERIOPHAGE

MEASLES, MUMPS

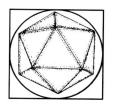

VIRUSES

iruses, the smallest known reproducing beings, are infamous for their role in causing colds, measles, mumps, cold sores, influenza, polio, smallpox, hepatitis, warts, and, at least in certain animals, cancer. Viruses were first discovered in 1917 when the French-Canadian microbiologist Felix d'Herelle, working at the Institute Pasteur in Paris and later at the Université de Montreal, found them infecting bacteria. D'Herelle called them "bacteriophages," literally meaning bacteria-eaters. D'Herelle argued for the existence of such entities when he discovered that some agent of reproduction had passed through filters with pores smaller than any bacterium.

Two years earlier, a British bacteriologist by the name of F.W. Twort noted an infection on his *Staphylococcus* bacteria that could be perpetuated and that drastically altered the appearance of the bacterial colonies. He published one brilliant short paper on the subject before he was conscripted into the British army and his studies came to an end. Filterable subvisible particles capable of causing indefinite infections in animals and plants had already been known and named viruses or virus pathogens. However, not until the bacterial viruses of Twort and d'Herelle was their essential nature—that they need a host cell to reproduce—really understood.

Today viruses (Figure 9.0), some so small that thousands of them can fit in the nucleus of a eukaryotic cell, can be seen, but only with the electron microscope. As d'Herelle claimed in his classic book, *La Bacteriophage: Son Role Dans D'Immunité*, written in 1921, viruses are tiny inheritable replicating entities. But viruses are not cells, nor are they capable of reproduction outside of cells. They are a strand or strands of DNA or RNA wrapped up in a coat of protein and sometimes in an outer protein structure as well. Whereas a human cell may consist of 10,000 genes that contain biochemical instructions for specific proteins, viruses exist with as few as four genes. They do not "live" at all, but remain outside the cell in a waiting stage, performing none of the functions of the self-maintaining cell. Yet while they do not maintain themselves, viruses do perpetuate themselves. They remain dormant, ready to plug into a cell's reproductive machinery, subvert it, and, as a sort of genetic automata, make more of themselves.

Once inside a cell, the DNA of viruses may be transcribed into RNA, which, instead of making normal cell proteins, makes viral ones, such as those

101

of the virus's protein coat. Other viral proteins may interfere with the host cell reproductive machinery so dramatically that the cell becomes ravaged, subordinated to the self-serving systems of the nonliving virus. In cells with nuclei, viruses use energy from the mitochondria and rearrange the invaded cells' amino acids into different proteins. Some of these proteins are enzymes that break and splice pieces of the host viruses into new virus protein and DNA. In a matter of hours, hordes of new virus particles assemble within and from the spoils of the taken-over cell. Some viruses break out, rupturing and destroying the invaded cell. These are known as the lytic viruses.

"Smarter" viruses seep out of the cell membrane. Because viruses depend on cells to reproduce but also can destroy them, the most successful viruses are ones that do not "bite the hand that feeds them "—in other words, ones that reproduce without completely destroying their intracellular environment. From this angle, less harmful viruses are better adapted than more harmful ones. However, a "safe" virus can turn into a deadly one when its cellular environment changes. For example, 30 different kinds of gland or adenoviruses are commonly found in human beings, and the only sort of problem they cause for certain is minor or temporary respiratory ailments. These same viruses, however, can cause chronic cancers when injected into rodents.

In addition to DNA viruses are the still stranger RNA viruses—structures that also may wreak severe cellular havoc but that, unlike DNA viruses, have no true genes of their own! Instead, RNA viruses are minute chemical bundles that mimic the host's messenger RNA, which turns DNA into proteins in living cells. They work by intercepting cell reproduction at the protein-building level like a disguised man who murders the foreman of a construction team and directs the unwitting workers to build a completely different structure in its place. RNA viruses are responsible for mumps, AIDS, yellow fever, Colorado tick fever, and foot-and-mouth disease of cattle to name but a few. The Bittner virus, which causes cancer in mice, is an RNA virus that is transmitted through the mother's milk, which may contain 50 billion viral particles per drop! Female mice suckled on such carcinogenic milk usually develop cancer, but not until they mature.

Oncogenic, or tumor-forming viruses first were discovered as the cause of leukemia in chickens. They work by invading cells and inducing them to reproduce more quickly, thus accounting for the cancerous tumor. In the initial stage, most of the cells invaded by a tumor-forming virus quickly die. But the ones that survive have their genes overhauled and changed from those of normal tissue cells into cancerous ones. These genetically redesigned cancer cells, samples of which have reproduced continually in laboratory cultures for over 20 years, are potentially immortal. When introduced into the appropriate tissue of the correct animal species, they again cause cancer by genetically changing the host cells. Oncogenic viruses fundamentally differ from other disease-causing viruses in that they do not so much destroy cells as exert control over their internal processes of differentiation, thus converting cells from whatever they were before into a new, faster-reproducing cell type.

Herpes viruses present the best evidence so far in linking cancers to viruses. Beside the statistical correlation linking genital herpes (herpes simplex II) with cervical cancer (most women who develop cervical cancer are infected with the virus), normal human tissue cells grown in culture become cancerous when adulterated with herpes simplex I, which causes cold sores and cankers in the mouth, or genital herpes. Since the wives of men whose former wives develop cancer of the cervix are three to four times more likely to develop cervical carcinoma themselves, men are thought to transmit a cancer-causing virus, possibly a form of herpes simplex II, from woman to woman. Sexual promiscuity has been accompanied by the diagnosis of cervical cancer in women at earlier ages, as well as by an increase in the number of genital herpes cases worldwide. A further unsettling correlation is that cancer of the penis is more frequent in men whose wives have or will develop cervical cancer.

Despite their frightening effects and correlation with cancer and acquired immune deficiency syndrome (AIDS), viruses are best thought of as random bits of genetic information that can be transmitted into live cells under certain conditions. Though they can change and destroy cells, they can no more function without the cells they inhabit than can a computer program exist without a computer. Indeed, a good analogy for viruses is the concept of a reproducing computer program. Imagine a planet of reproducing computers. The computers duplicate themselves by reproducing master programs which direct automata to make more computers and more computer programs. The computers are analogous to normal cells. Now come some computer programs sufficiently complex to introduce themselves into master programs and direct the automata to make different machines. These machines are like non-oncogenic viruses. Now imagine a computer program with the information necessary to change the master programs and induce the computer to speed its production of new computers that also contain the new master programs. These new master programs are like the oncogenic viruses, and the converted computers are analogous to cancer cells. As aliens visiting such a computer planet we might admire the deceptive prowess of the converting master programs until, of course, some one points out that such programs are like the viruses that cause cancers.

Viruses are too small to be seen by you; an electron microscope is needed. The best way for you to see a virus is to find a picture of one in a book on the subject. Vital plaques, evidence that viruses are present in a bacterial culture are simply holes made on a petri plate by the lytic bursting of bacteria. Because viruses do not move, the plaques don't actively spread. Instead they are passively moved by bacterial hosts coming in contact with one another. Because plaques do not grow larger than a given size characteristic of the virus in question, viruses can often be told apart with the naked eye. The curious naturalist can see viral plaques by visiting a virology lab.

FIGURE 10.0

Bacterial shapes: (top to bottom) vibrio, spirillum, spirochete, bacillus, and six cocci. Drawing by J. Steven Alexander.)

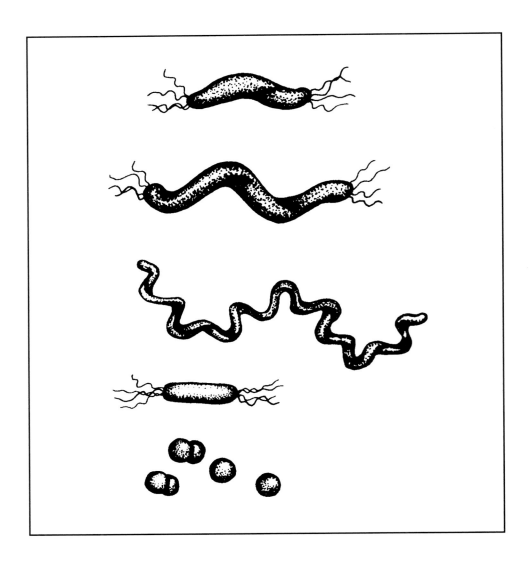

T·E·N

BACTERIA

■ INTRODUCTION TO BACTERIA ■

s Anthony van Leeuwenhoek realized in the 1600s, bacteria are everywhere there is life. In rich, complex, still largely unknown mixtures, bacteria grow and in general live more comfortably than protoctists or fungi. Together bacteria are chemically more versatile and capable of surviving harsher "insults" than eukaryotes. In environments within bounds tolerated by life, bacteria can take greater heat, greater cold, more acidity, more alkalinity, more pressure, more desiccation, and so forth than eukaryotes. They live inside rocks, between sand grains in the desert, in scalding springs, and in the stratosphere. Some 20,000 different kinds of bacteria have been estimated to exist but, because they exchange genes so readily, there are probably far more to be reckoned with. Indeed, Clair Folsome of the University of Hawaii has remarked that there are as many kinds of bacteria as there are bacteria.

The kingdom Monera (Figure 10.0), from the Greek *moneres*, meaning single or solitary, is a synonym for the kingdom of bacteria. Although as "germs" bacteria are responsible for many diseases, far more are neutral or positively helpful in their associations. They are crucial to the health of forests, people, and the planet as a whole. For example, one spoonful of high quality soil contains about 10 trillion bacteria. Gardens, crops, and trees growing in such soil feed farm animals and humans, which are returned to ecologically usable form in the bacteria-aided process of decay after death.

Unlike members of the other four kingdoms, which are limited in their chemical repertoires, bacteria emit and incorporate all the major reactive gases of the atmosphere. Bacteria are part of a worldwide circulatory system influencing the balance of nitrogen, nitrogen oxides, oxygen, carbon dioxide, carbon monoxide, hydrogen, methane, ammonia, and sulfur gases in the atmosphere. Without our continually modulated, chemically unstable atmosphere, so-called "higher" organisms—the plants, animals, fungi, and protoctists—would never have evolved.

Because of their fluid gene exchange and unrivaled metabolic diversity, bacterial classification is never absolute—as with anything on an evolving planet. The classification of bacteria used in this Guide differs from books

105

such as *Bergey's Manual of Determinative Bacteriology* in that we center upon the taxonomic level of phylum (e.g., anaerobic sun lovers, gas eaters). One should have a basic distrust of the concept of species when applied to bacteria. Because all bacteria potentially can trade genes with all others, and because a species is defined as a group of organisms that only breed with each other, bacteria are not different species so much as a single spatially separated but genetically unified super-species.

Bacteria are identified in this Guide by a combination of metabolic and structural traits. As much as possible, when similarities have probable evolutionary connections we will treat them together under the same heading. In our Guide we discuss only the major, best-known, bacterial groups. In the final analysis, however, the groupings are only rough perceptual guidelines. Bacteria can evolve, spread genes, and adapt so fast that no possible human taxonomy can truly contain them.

▪ WALL-LESS BACTERIA ▪

The wall-less bacteria (aphragmabacteria) are unique in the bacterial world: they lack the ability to produce the rigid cell walls typical of all other known bacteria. This "weakness" becomes a strength when these wall-less organisms come up against such antibiotics as penicillin, which work by inhibiting the growth of bacterial cell walls. These antibiotics have no effect on the wall-less bacteria and, through gene trading, aphragmabacteria may confer their powerful defense system to other types of previously vulnerable bacteria, making them impervious to antibiotic action as well.

The different varieties of aphragmabacteria may not be directly related to each other; they may simply share the trait of wall-lessness. On the other hand, some wall-less bacterial strains may be directly related to each other and to the earliest bacteria that never evolved cell walls. Evidence for this connection is that most wall-less bacterial cells contain a very small quantity of DNA.

The naturalist can find wall-less bacteria in a variety of habitats, from the leaves of citrus plants (where some cause the disease called "stubborn") to the bodies of many types of plants and animals. Most of the well-studied aphragmabacteria have been implicated in disease production. The most well-known genus, *Mycoplasma* (Figure 10.1), is notorious for its role in several kinds of pneumonia known to infect both humans and domestic animals. *Mycoplasma* also routinely infects laboratory tissue cultures of human and other cells.

Chemically, mycoplasmas are oddities in the bacterial world in that more than 33% of their unusual cell membrane is composed of cholesterol. Cholesterol is commonplace in animal tissue but such high proportions are unknown in other bacteria. Today all mycoplasmas taken from plant or animal hosts and grown in the laboratory require steroids, such as cholesterol, to survive. Perhaps the long association between certain mycoplasmas and animal tissue provided the opportunity for *Mycoplasma* to sequester materials made by the animal cells for their own growth and reproduction.

Wall-less bacteria survive in some of the harshest environments on earth—the microcosmic equivalents to Death Valley or the Sahara Desert.

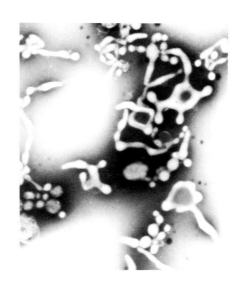

FIGURE 10.1

Mycoplasma, a wall-less bacterium (phosphotungstic acid-stained specimen seen with an electron microscope).

106

Thermoplasma is a genus of aphragmabacteria containing a single species. It thrives in extremely hot and acidic conditions where the temperature hovers near 140°F (60°C) and the level of acidity ranges from a pH of one to two—the acidity of *concentrated* sulfuric acid. In fact, *Thermoplasma* is difficult to study because it freezes to death when the temperature drops below 100°F. Because they freeze when put on a glass slide at room temperature, no one yet knows what *Thermoplasma* look like alive! The only colonies found so far have come from the surface of hot piles of coal and the scorching springs of Yellowstone National park, so the amateur naturalist would be very lucky to discover *T. acidophilum.*

Thermoplasma is fascinating for another reason: it is the only prokaryote to contain proteins similar to histones, which are tangled up with the DNA of chromosomes in plants and animals. Of all the aphragmabacteria, *Thermoplasma* is the one most similar, in terms of its RNA, to methanogenic and certain salt-loving bacteria that are sometimes called archaebacteria, or "old bacteria." Because of this similarity, *Thermoplasma* also is sometimes called an archaebacterium. All three archaebacteria groups are distinguishable from eubacteria, or "true bacteria," by certain chemical differences in their RNA and lipid–ether linkages. Indeed, they may be possible biochemical "missing links" between bacteria and all the more complex forms of life made of nucleated cells, from amebas and paramecia to redwoods and whales. A distinct evolutionary possibility is that *Thermoplasma* bacteria, or their ancestors, were symbiotic prey to other, predatory bacteria that did not succeed in killing them but eventually survived with them. The predators could have come to live inside *Thermoplasma*, providing them with oxygen-derived energy. In any case, the presence of histones in both *Thermoplasma* and more familiar, larger forms of life suggests an ancient evolutionary connection between these bacteria and the cells of which we are made.

Unfortunately, without experience and expensive apparatus the wall-less bacteria can be collected only from live plant and animal hosts. Although they are present in many plants and even insects, most aphragmabacteria are so small they cannot be identified easily, even with powerful electron microscopes. The determined naturalist, however, may achieve a measure of success: wall-less bacteria often form tiny but recognizable colonies in nutrient cultures on petri plates. Like a fried egg, a dark or colored "yolk" will be surrounded by a lighter "white" area (Figure 10.2.).

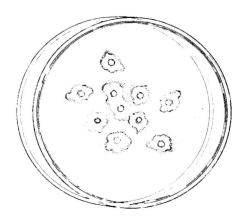

FIGURE 10.2
"Fried egg" morphology of *Mycoplasma* colonies growing on the surface of agar in a petri plate. (Drawing by J. Steven Alexander.)

▪ FERMENTERS ▪

The fermenters, or fermenting bacteria, date back to the time before major amounts of oxygen accumulated in Earth's atmosphere. Today they are found, and indeed proliferate, *under* the earth—in soils, river beds, marine sediments—wherever the oxygen they find so poisonous can be avoided. Our life-sustaining gas is to them a pollutant which, if it does not always kill them, at least stunts and blocks their growth.

Fermentation is a process in which organic compounds—the complex, hydrogen-rich carbon compounds associated with all living things—are used both as food and to extract chemical energy. Gram negative cocci, for example, may ferment carbohydrates into carbon dioxide; in doing so they

107

derive the energy needed for cell growth. Lactic acid bacteria, found in spoiling milk, ferment milk sugar (lactose) into lactic acid. Perhaps the most familiar examples of fermentation are the many active fermenters that metabolize the sugars in grape juice into ethanol (ethyl alcohol), the agent of inebriation in wine and other alcoholic beverages.

The first form of life on Earth may well have been a kind of fermenting bacteria. In the planet's earliest days, the Sun, through a process called "prebiotic synthesis," is thought to have produced many megatons of the types of organic compounds transformed by fermenters. Lightning, another energy source more common during the time of Earth's turbulent birth from materials in space some 4.6 billion years ago, also would have aided the inanimate production of compounds needed by ancient forms of fermenting life.

Much later, fermenters migrated to the guts and mouths of animals-oxygen-free habitats that arose only with the evolutionary development of animals themselves sometime after 600 million years ago. Because the inside of an animal is an environment enticingly free of gaseous oxygen, some fermenters, such as *Clostridium botulinum*, actually prefer to grow inside animal tissues where they can contribute to or cause such diseases as botulism and gangrene. Luckily for animals, however, most fermenters, like those that make acetate from sugars in the guts of cows and termites, are either benign or are even helpful to the animals in which they live.

You can collect fermenting bacteria in a variety of locales. A glass of milk left for 24 hours at room temperature makes a fine "trap" for fermenting bacteria. Unfortunately, such a rich food source attracts many other types of airborne bacteria and fungi as well. In order to avoid attracting nonfermenters, proceed as follows: Take two small sterile cans of evaporated milk, pour out half of each can and dilute each with a portion of boiled or distilled water. To one can add a spoonful of yogurt, one that says "active cultures" on the label. Add nothing to the other, which will serve as the control. The yogurt you have added, containing active cultures of bacteria that will ferment the sugar in milk to lactic acid, serves as an inoculum. (Indeed, this is basically the method used in making commercial yogurt.)

Cover both cans and leave them to sit at room temperature or warmer for several hours or more. Upon examination you will likely notice two kinds of fermenters: *Streptococcus*, which grows in chains of circular cells, and *Lactobacillus* (Figure 10.3), a rod-shaped bacterium. These pure cultures can be compared to the mixed ecology of the microbes in a glass of milk which has been left out, closed or uncovered, for a day or longer.

The typical nose-wrinkling smell of spoiled milk indicates microbial activity, usually active fermentation by such successful bacteria as *Lactobaccillus*, *Streptococcus*, and *Leuconostoc*. While sour milk is no culinary delight to humans, the fermenters feast, turning milk sugar not only into lactic acid but also into alcohol and other chemical compounds. Carbon dioxide in small quantities of 5 to 10% (which probably mimic the condition on the early Earth) enhances fermentation. Add a shot or two of soda water or sodium bicarbonate, the "fizz" of which is carbon dioxide gas, to souring milk or yogurt cultures as an experiment. Fermenters stain Gram positive and high acidity is often a good indicator of their presence.

If we are not familiar with *Lactobacillus*, it is certainly familiar with us! Bits of food with the formic acid and high concentrations of B vitamins found

108

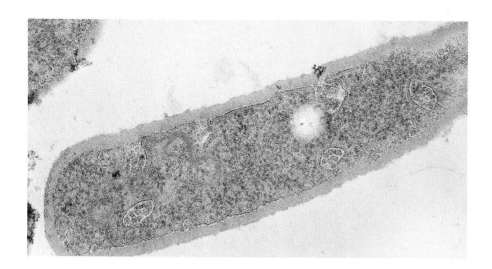

FIGURE 10.3
Lactobacillus, a fermenting bacterium. (Electron micrograph courtesy of David Chase.)

in our saliva make a good medium for the growth of *Lactobacillus,* which is normally found among the estimated 50 billion microbes that inhabit the human mouth.

While some fermenters, such as the clostridia, have the ability to form heat or desiccation-resistant spores that allow them to survive in the oxygen-rich air of the Earth's surface, milk fermenters have no such ability. These "bugs"—as microbiologists affectionately call microbes—already exist in pasteurized milk, although in sparse numbers. They would be poisoned by the oxygen if left in the open air.

Fermenting bacteria also are responsible for a wide variety of fine wines, cheeses, and liquor. But their commercial importance stretches beyond the worldwide production of culinary delicacies. Fermentation also results in the industrial production of nutrients and vitamins such as citric acid, riboflavin (vitamin B_2), vitamin B_{12}, and antibiotics. The Japanese even exploit fermenting bacteria to produce L-glutamic acid, the salt that is the source of monosodium glutamate, the widely used food flavoring (its name, in Japanese, means "source of taste"). Fermenting bacteria may be even more important in the future because they hold the promise of evolving into microorganisms that can break down complex organic compounds, such as those found in oil spills and new plastics, which pose an increasing and dangerous risk in our human environment.

▪ SULFATE REDUCERS ▪

Although the sulfate reducers are poisoned by oxygen, they "breathe" sulfate in much the same manner that we breathe oxygen. Chemically, breathing or respiration is a process that yields energy through the oxidation of food molecules to inorganic compounds such as H_2O, CO_2 or sulfate. In the process, the oxygen becomes richer in hydrogen (that is, it becomes reduced). Sulfate reducers technically belong to the phylum Thiopneutes, which, not surprisingly, means "sulfur breathers" in Greek. Besides sulfate, members of this phylum take in organic carbon compounds and some, such as certain marine forms, require sodium chloride

FIGURE 10.4

Desulfovibrio, a kind of sulfate-reducing bacteria. (Drawing by J. Steven Alexander.)

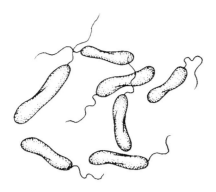

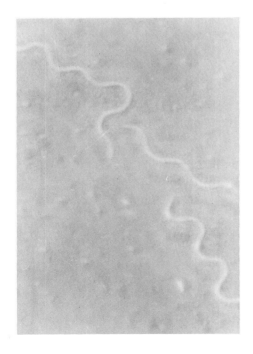

FIGURE 10.5

Pillotina spirochetes—the largest and most complex spirochetes known. All known pillotinas come from hindguts of wood-eating cockroaches and termites.

(table salt) for growth. By chemically turning oxygen-rich sulfate into sulfur or hydrogen-rich sulfide, sulfate reducers provide a valuable service to organisms that require sulfur but cannot use it in its oxidized form.

The three main genera of sulfate-reducing bacteria are *Desulfovibrio* (Figure 10.4), *Desulfotomaculura*, and *Desulfuromonas*. Members of these genera are widely distributed in nature. For example, *Desulfovibrio* may be found stinking up marine and freshwater muds and brackish estuaries. Because sea water has much more sulfate than fresh water, these organisms are best known to marine microbiologists. *Desulfotomaculum* can be captured from muds and soils as well as from some spoiling foods and the guts of insects and bovine animals. These unicellular bacteria move by means of many flagella waving at the cellular surface.

Sulfate reducers may be collected from seaside muds or swampy areas that give off a foul odor reminiscent of rotten eggs or industrial settings such as parts of the New Jersey Turnpike. This smell is hydrogen sulfide—a gas so poisonous to people it can kill us before we smell it, yet, due to a strange quirk of olfaction, so smelly it can be detected in parts per million. Hydrogen sulfide is produced when *Desulfovibrio* or *Desulfotomaculum* inhale the oxidized forms of sulfur, namely sulfate or sulfite. They then reduce the compounds by donating electrons from the hydrogen atoms of food. The result is the hydrogen-rich, gaseous compound hydrogen sulfide. Little bubbles of the gas accumulate, rise, and pop up at the surface of various types of muck, where they release that familiar mephitic stench.

While sulfate reducers have few direct commercial applications, their value to the global ecosystem—the ultimate supporter of all commercial enterprises—is inestimable. By turning sulfate and sulfite, which are produced by other microbes as waste, back into sulfur and hydrogen sulfide, the sulfate reducers keep the element available. Continually in limited supply, sulfur is one of the most vital resources in the biosphere. Since the proteins of all cells in all organisms contain sulfur, sulfate reducers form a crucial part of the global sulfur cycle. In addition, iron sulfide—mineralogically known as iron pyrite—may be rock-solid evidence of ancient ecosystems rich in sulfate-reducing bacteria. The reason: when sulfate reducers produce hydrogen sulfide in iron-rich waters, the gas reacts with the iron to form the iron sulfide, or "fool's gold," found in ancient rock formations.

For true nature lovers able to stand a bit of knee-high muckraking, hunting sulfate reducers will prove exciting indeed. They are often found in malodorous muds and in intertidal areas somewhat removed from the seashore such as can be found, for instance, in the salt marshes lining the Atlantic Coast or those around the San Francisco Bay.

▪ SPIROCHETES ▪

Spirochetes (see Figures 10.5 and 2.2, 2.3) are incessantly wriggling, exceedingly thin, corkscrew-shaped bacteria found in places rich with organic materials. Spirochetes were first seen by Leeuwenhoek in the late 17th century. He observed them by taking samples of his own excrement during a period of ill health and of scrapings from his gums and the back of his molar teeth. Leeuwenhoek's observations are easy to repeat. Scrape some material from your back teeth with a toothpick and put it on a glass microscope slide. You will need to use a magnification of at least 400 to clearly distinguish the spirochetes from spirilli.

Spirochetes, by definition, are bacteria that bear internal flagella. These so-called periplasmic flagella are unlike the flagella of all other bacteria, which typically protrude into the medium. Flagella in spirilli, for example, poke through holes in the outer cell membrane. Furthermore, spirochetes tend to be more flexible, more wriggly, skinnier, and less defined in shape than spirilli. If you were the size of a microbe you would perceive spirochetes as little water snakes which, lacking heads or tails, whip back and forth with rapid darting movements in both directions.

Spirochetes like company and places with unpleasant odors. They inhabit organic films on puddles and ponds, and sulfurous muds and stenchy piles of decaying algae. They can be found in enormous numbers in the rumens (special grass-digesting stomachs) of cattle and in the hindguts of such insects as wood-eating cockroaches and termites. One of the best places to find spirochetes in large numbers is in clams or oysters.

Just before you are ready to search for spirochetes (and you should plan to spend an afternoon at it) collect fresh, edible clams or oysters. Open them with a clam knife and gently poke around the mantle cavity. Identify the glassy, upside-down cone organ called the "crystalline style" toward the front end (Figure 10.6). This is an organ of the digestive system of bivalve mollusks such as clams, scallops, and oysters. When it has the consistency of firm jello it tends to be packed with hundreds of writhing, snake-like spirochetes called "cristispires." As the styles become too hard, like butter taken directly from the refrigerator, or too soft, like warm Jell-o, spirochetes rapidly swim away looking for styles of appropriate consistency. Unlike other bacteria, spirochetes are extremely effective at swimming rapidly through dense or viscous media. For this reason their favorite places are firm muds,

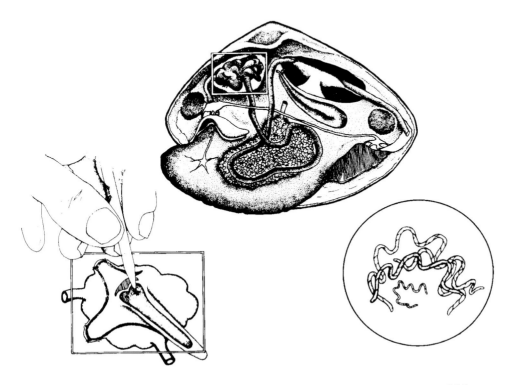

FIGURE 10.6

Spirochetes in the crystalline "style" of a clam. To see these, a drop of fluid is taken from the anterior stomach of an oyster or clam (where the "style" is located) and placed on a microscope slide. These spirochetes are so big they can be seen at 400x with a light microscope. (Top and left drawings by Sheila Manion-Artz, right drawing by Christie Lyons.)

111

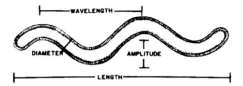

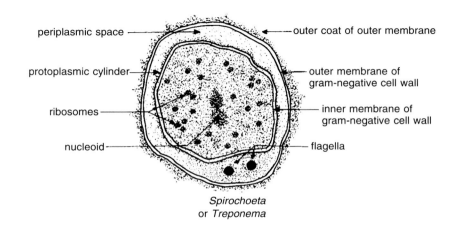

Spirochoeta
or *Treponema*

FIGURE 10.7

Characteristic detailed structure of a spirochete as reconstructed from transmission electron micrographs. (Above drawing by Laszlo Meszoly, drawing at right by Barbara Dorritie.)

FIGURE 10.8

Pillotina spirochete from the hindgut of a termite (transmission electron micrograph cross section). The wheel-shaped bodies at the top and left of this micrograph are cross sections of undulipodia of other microbes (protists) in the hindgut community. (Electron micrograph courtesy of David Chase.)

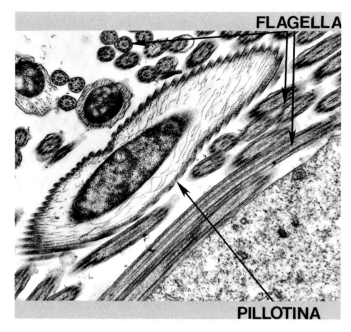

testicular tissue of animals, piles of decaying vegetation, the firm crystalline styles of bivalve mollusks, and rumen fluid.

While the vast majority of spirochetes are not known to scientists, people intent on finding them have no trouble if they look in the right places. In fact, the most famous of all spirochetes, *Treponema pallidum*, is found in huge numbers associated with the skin sores of people with symptoms of the venereal disease syphilis. In late stages of untreated syphilis, years after the first sores are noted, wriggling spirochetes are often found attached to brain tissue of afflicted people.

Even though oxygen-requiring (aerobic) spirochetes have been found, most are strict anaerobes. In addition, no one has ever discovered any spirochete able to live by photosynthesis. Thus, spirochetes often are found in such oxygenless environments as muds, which harbor brightly colored photosynthetic bacteria.

The elements of spirochete structure include the protoplasmic cylinder, the periplasmic flagella, the insertion rotary motors, and the helical body with

112

characteristic wavelength and amplitude (Figure 10. 7). In Figure 10.8, you can see the giant termite spirochete, *Pillotina*, which, with its crenulations (ruffles) and sillon (deep groove) and over 100 periplasmic flagella, is the most complex spirochete yet discovered. Unfortunately, no one has been able to coax *Pillotina* to grow outside the hind end of the various drywood termites in which it is always found.

▪ METHANOGENS ▪

Methanogens are highly unusual bacteria that produce methane. Like the wall-less *Thermoplasma*, they are also called archaebacteria. The so-called archaebacteria share traits with each other that are distinct from all other bacteria. Collectively, methanogens act as a chemical beacon illuminating the existence of carbon-based life on Earth. We say this because methanogens produce so much methane ("swamp gas") that it could easily be detected by atmospheric chemists living elsewhere in the solar system. Methane reacts rapidly and completely in the presence of oxygen to form carbon dioxide and water. The continuous presence of significant amounts of methane in an oxygenated atmosphere means that it is continuously being manufactured. Its manufacturers are the methanogens.

Methanogens grow in sewage, sediments, and swamps all over the planet. They are also found in great abundance in the intestines of all large animals, including humans. Above all, methanogens are found in the guts of animals with rumen stomachs, such as cows, which have been called "forty-gallon methane tanks on legs" (Figure 10.9). An estimated 30% of atmospheric methane comes by way of grazing animals such as goats, horses, and cows. This has led to the semiserious suggestion that the primary function served by large mammals is the equitable distribution of methane gas throughout the biosphere.

The methane the bacteria emit builds up in sewage-treatment facilities and is the source of the natural gas used to heat houses and stoves. (Because natural methane has no odor to warn us of its flammable existence, it is commercial practice to add noxious-smelling fumes before the gas makes its way through pipelines across continents into suburban kitchens.)

The physiology of methanogenic bacteria, technically belonging to the phylum Methanocreatrices, is bizarre even for bacteria. As we have said, some biologists have suggested that the methanogens along with certain other obscure bacteria, including hot-acid-tolerant organisms such as *Thermoplasma* and *Sulfolobus*, and the salt-loving halobacteria (Figure 10.10), should form a separate group of archaebacteria; some even think these organisms are so different they represent a whole new kingdom of living organisms. The primary reason for keeping the methanogen group separate lies in their RNA. The RNA of the ribosomes of the methanogen group is so different from that of most other bacteria that molecular biologists believe these forms of bacteria have an ancient heritage and have evolved along their own separate path for billions of years.

Methanogens do not use sugar, proteins, or carbohydrates as sources for food and energy. The most parsimonious eaters around, they live off pure air. They simultaneously oxidize hydrogen or certain hydrogen-rich chemicals

FIGURE 10.9

"A 40-gallon methane tank on four legs." Methanogenic bacteria, such as those shown, live in the rumen of cows and other grazing animals and contribute significant amounts of methane gas to the atmosphere. Names are those of various types of methane-producing bacteria. (Drawing by Louane Hann.)

FIGURE 10.10

Halococcus, a salt-loving coccoid bacterium. The rectangular object with a halo in this light micrograph is a crystal of ordinary table salt (sodium chloride).

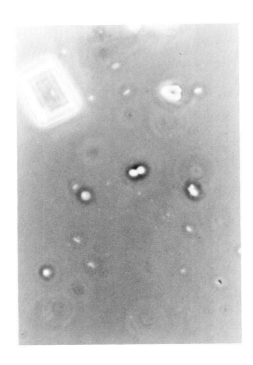

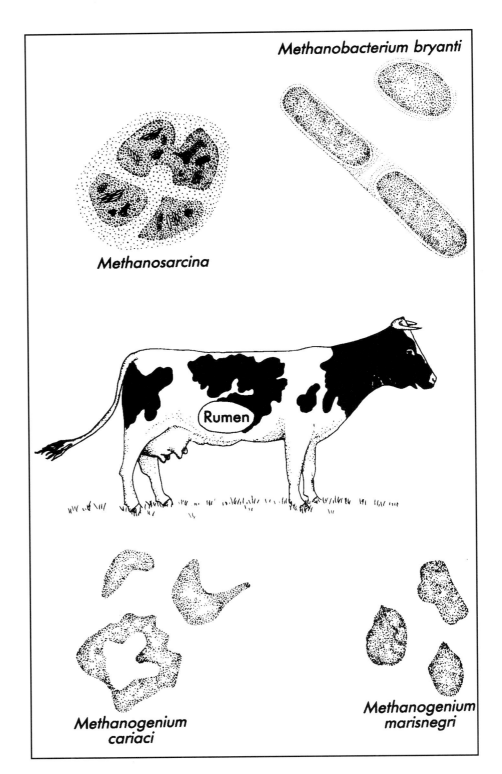

and chemically reduce atmospheric carbon dioxide. The resulting waste consists of methane and water. Most of the methane in the Earth's atmosphere—1.5 parts per million of the air we breathe—is produced by methanogens, which emit a staggering two billion tons of this natural gas per year.

Reactive methane emitted by methanogens is vital to the health of the biosphere as a whole. If methanogens did not remove carbon from sediments

114

and return it as methane to the atmosphere where it reacts to form CO_2, carbon buried in the ground would pile up in huge deposits because of photosynthesis by blue-green bacteria, algae, and plants. Excessive amounts of oxygen would accumulate in the atmosphere. The result would be worldwide conflagrations and terrible natural firestorms due to too much oxygen. We do not live with the threat of such disastrous firestorms in part because the natural gas produced by methanogens reacts with oxygen to form more carbon dioxide, thus keeping oxygen levels in check.

Since hydrogen-rich gases like methane are reactive in an oxygen-rich atmosphere, energy in the waste of methanogens can be put to use by humans. Methanogens are economically important as the major cleaning elements in sewage-treatment plants. And again, they are directly or indirectly the main providers of natural gas—an abundant, low-cost, and clean form of fuel used in the manufacture of such diverse products as carbon black, methanol, formaldehyde, chloroform, carbon tetrachloride (a cleaning fluid), and nitromethane.

Although methanogens are anaerobic bacteria and, therefore, must live in waters or muds in nooks and crannies where oxygen is excluded, the gaseous emissions of methanogenic bacteria would make it easy for extraterrestrials to detect life on our planet from space. Given the fact that our atmosphere is 21% free oxygen, methane should not be found in it at all. Nonetheless, our air contains more than one ten-thousandth of one percent methane. By the rules of equilibrium chemistry—the science of how chemicals react to form a steady mixture—even this small amount is far, far more than is expected; it exists because it is being produced far faster than it can react with oxygen. Extraterrestrial scientists, if they could measure this atmospheric oddity, might be able to tell there was life on Earth from millions of miles away.

Methanogens are extremely difficult to grow by themselves in culture. They may, however, be collected from swamps and marshes. About 10 drops of sediment suspected to be rich in methanogens can be placed in a small glass tube in which oxygen has been removed. To remove oxygen, stuff a damp soil sample into a tightly closed tube for a week or so, then take out the soil and fill the tube up to the top with your water sample. If their favorite gases, hydrogen and carbon dioxide, are not available, methanogens will eat acetate, methanol (wood alcohol), or trimethylamine. If a rubber membrane—a balloon will do—is placed across the ends of the tube it will swell up as gas is produced in the dark. The gas can be assumed to be methane, which burns, torchlike, when a lighted match is put to it.

■ SUN-LOVING ANAEROBES ■

In the history of life on the planet Earth, nothing has had so many repercussions throughout the ages as the development of photosynthesis. This giant step occurred not in plants but in tiny photosynthetic anaerobic bacteria (Figures 10.11, 10.12). At least four types of sun-loving anaerobes still abound on earth: the green sulfur, the purple sulfur, and the purple nonsulfur bacteria, and the brownish nonsulfur *Heliobacterium*. Because all four types adapted to the ancient, oxygenless environment of Earth, today they all are hampered or destroyed by ambient oxygen.

115

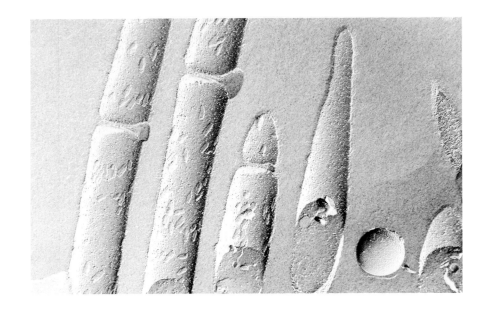

FIGURE 10.12

Rhodospirillurn rubrum, a purple-sulfur anaerobic phototrophic bacterium. The spheroidal inclusions shown in this transmission electron micrograph are granules of the storage product—β-polyhydroxyburyrate. (Photo courtesy of Germaine Cohen-Bazire.)

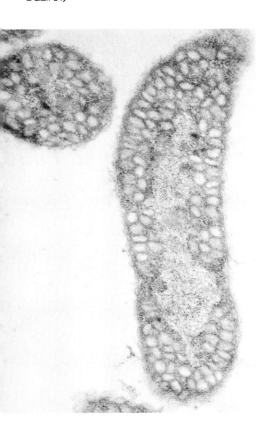

Photosynthetic beings require both light and carbon. Their carbon source almost always is atmospheric carbon dioxide; if not, it is some small carbon compound. But they also must have a source of electrons, which in practice means a source of hydrogen. It is in their sources of hydrogen that anaerobic photosynthetic bacteria and plants differ. Whereas plants get their hydrogen from the H_2O of water, photosynthetic anaerobes use other, more chemically accessible hydrogen sources that were prevalent on the early Earth during their development some three billion years ago. Hydrogen sulfide, originally emitted from the earth's interior through volcanos and submarine cracks in the Earth's crust, was one early source. (Today hydrogen sulfide is formed by many bacteria, including the sulfate reducers.) A still more ancient source of hydrogen tapped into by the anaerobic photosynthesizers was hydrogen gas (H_2) itself.

Hydrogen gas probably was much more available on Earth billions of years ago than it is today. Stars, formed from collecting clouds of hydrogen gas, eventually ignite in a process of nuclear fusion that pushes leftover hydrogen away into the deep reaches of space. The early Sun, less bright than it is now, had not yet pushed hydrogen gas away from our planet into the outer solar system, where it has collected around the larger planets Jupiter and Saturn due to their immense gravity. Early photosynthetic forms of life used such hydrogen while it was still available.

The anaerobic photosynthesizers today thrive chiefly in such oxygenless environments as waterlogged soil, lake and puddle scums, the edges of slow rivers, and on the seaside expanses that flood with the tides. The entire layered community and the sediment particles trapped by it are known as microbial mats. The mats consist of richly hued pink and green layers varied in texture and living beings. The photosynthetic anaerobes inhabit the lower strata of microbial mats where they can be safe from the harmful effects of surface oxygen. Lower layers of a microbial mat (which, due to the reaction of iron with sulfide may be conspicuously black on the top) are often rich in purple and green tones, reflecting the presence of hydrogen sulfide-using purple sulfur bacteria, purple nonsulfur bacteria, and green sulfur bacteria.

The best way to collect photosynthetic bacteria is to be on the lookout for purple, red, and bright brownish or greenish colors as scums on ponds or puddles, or as thick mats in such areas as the sandy zones between salt marsh grasses. A quart bottle filled to the top with sand and mud from the seashore and placed in a brightly sunlit place at room temperature or warmer will develop an excellent community of photosynthetic bacteria in a few weeks. Add a shredded paper towel (or cellulose fibers) and a spoonful of calcium sulfate: bacterial growth will be luxuriant. A comparable "control" bottle placed in the dark may reveal microbial communities but not of photosynthetic bacteria, which require light. This setup is basically a homemade "Winogradsky column," similar to those devised by the Russian scientist Sergei N. Winogradsky, who emigrated to the Institute Pasteur in Bric-Compte-Rokett, France, in the 1920s.

Heliobacterium, a photosynthetic bacterium with a special type of bacterial chlorophyll, was discovered as a brownish film on such a bottle left in the light and containing a sample of soil taken from near the microbiology department of the University of Indiana. An observant graduate student in the early 1980s was astute enough to track the growth of this new photosynthetic organism. Just as Winogradsky captured the environment of anaerobic photosynthetic sulfur bacteria, so you too can verify the persistence and changing patterns of ancient life trapped in a bottle. You will see the growth, demise, and replacement of varicolored communities as long as your bottle is exposed to the energizing effect of light.

Under the microscope, light-using anaerobes vary greatly in appearance. They may be unicellular or multicellular; if multicelled, they may form packets, stalked budding structures, or lines or sheets of cells. The spaces between the cells may be connected with a stringy mucous material called sheath. Some bob up and down with the aid of internal pockets of gas called vacuoles, which can be recognized because they tend to make the cell "sparkle." Tiny yellow granules—waste products of sulfur and sulfates—may be visible within the cells.

At present, anaerobic photosynthetic bacteria have little commercial or agricultural importance. However, their importance in the future in closed ecological systems orbiting in space may be crucial. This is because on Earth—the only complete ecological system known—the anaerobic photosynthetic bacteria and the sulfate reducers provide a vital link in the cycling of the limited biological resource, sulfur.

▪ BLUE-GREEN BACTERIA ▪

Although they evolved in massive numbers and spread to dominate the surface of the earth more than 2000 million years ago, blue-green bacteria (cyanobacteria) still are probably the most productive form of life on Earth. Hundreds of species of them exist (Figure 10.13). Their rampant growth and use of water as a source of hydrogen for photosynthesis led to a planetary change far more impressive than any wrought by man. In addition, because cyanobacteria lived symbiotically within other cells, which aggregated or never separated, blue-green bacteria evolved into the sun-loving plastids of all algae and plants. Seaweeds, algae, and waving

117

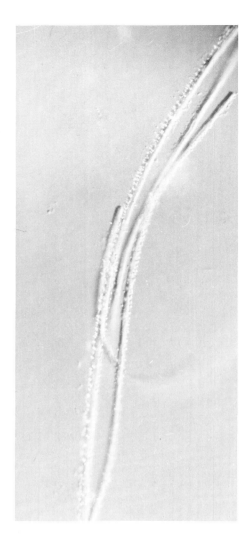

FIGURE 10.13

Oscillatoria limnetica, a cyano-bacterium with sulfur globules adhering to its outer surface. (Photo courtesy of Yehuda Cohen.)

fields of green grass may be photosynthetic bacteria that became superbly adapted to an intracellular habitat.

Until recently, cyanobacteria were considered simple marine plants. But they lack nuclei and have all the traits of bacteria, including penicillin-sensitive, non-cellulosic cell walls and the ability to host lytic bacteriophages, and so they must be considered bacteria. Some also have the non-plantlike trait of respiring only in the dark.

Blue-green bacteria literally changed the Earth: by exploiting water instead of hydrogen gas or hydrogen sulfide during photosynthesis, they tapped into a nearly inexhaustible hydrogen source. In the wake of this fortuitous exploitation, blue-green bacteria filled the atmosphere with their waste—oxygen. Today, we are completely dependent on the excretion of one of life's "lowest" forms; we breathe their waste in a deeply intimate ecological relationship.

The heyday of blue-green bacteria was during the Proterozoic Eon some 2500 to 600 million years ago, a time during which oxygen gas accumulated until it accounted for one fifth of the atmosphere. Many kinds of cyanobacterial communities survive today, including the spectacular domed, layered, hardened structures called stromatolites. Much smaller and more restricted than they were in the past, live stromatolites can be found growing in secluded places from the Persian Gulf and the Bahamas to the west coast of Mexico and western Australia. Fossil stromatolites are more common. They are thought to have been produced by many kinds of microbes, all of them living on photosynthate (food) produced mainly by filamentous kinds of cyanobacteria.

Due to their amazing reproductive success and ability to withstand harsh conditions, cyanobacteria are not difficult to find and collect. Like purple and green photosynthetic bacteria, cyanobacteria can be collected from colorful, textured seaside microbial mats. Cyanobacteria usually form the top, green layer in these mats. You also can collect them from the green scum atop swimming pools or polluted lakes and ponds. You can find them as the olive film covering seaside rocks. You can even scrape them off the walls of shower stalls and drains. Look for greenish tinges in damp, well-lit places. Blue-green bacteria are so well adapted to this planet that they have even been collected from the water used to cool down nuclear reactors. (Because cyanobacteria themselves, by expelling huge amounts of oxygen into the atmosphere, produced the radiation-absorbing "shield" of stratospheric ozone (O_3), they evolved prior to any atmospheric protection from radiation and to this day remain resistant to its effects.)

In addition, about 100 strains of blue-green bacteria are able to fix nitrogen from the air, freeing them from the necessity of living on soil. Thus, these organisms can invade and repopulate areas as barren as any on Earth. Three years after the Krakatao volcanic eruption wiped out all life on the surrounding island, cyanobacteria were growing on the volcanic ash and tuff. Indeed, cyanobacteria may be the single best-adapted life form on the planet today: they were the first form of life to settle the Bikini coral island after the nearby nuclear tests by the United States.

Cyanobacteria can be divided into two great classes: the coccoid, spherical forms, and the filamentous cyanobacteria. The round cyanobacteria are grouped into three major orders, mainly according to their mode of reproduction. Those that belong to the order Chroococcales, such as the stromatolite-building *Entophysalis*, reproduce by simple division. Some cyanobacteria reproduce by releasing smaller structures outside the cell called exospores or baeocytes. These are not really spores as they aren't resistant to anything; rather they are propagules, carriers of reproductive cells. Others reproduce by forming endospores—internal offspring propagules that destroy their parent when released. These do not resist heat or desiccation either.

In the filamentous class of cyanobacteria, multicelled filament fragments (hormogonia) break off and glide away to form new colonies (Figure 10.14). Some cyanobacteria show extremely complex forms, such as those in the group Stigonematales, which make filaments that branch. Some form the most complex and intricate structures in the entire prokaryote kingdom, while others form delicate iridescent green colonies capable of breaking away and growing new colonies like the *Gomphosphaeria* shown here (Figure 10.15). When nitrogen starved, the long *Anabaena* (Figure 10.16) cells give rise to round "heterocysts." Metabolically, heterocysts are non-photosynthetic and fix nitrogen, thereby solving the problem of nitrogen starvation.

Some species of cyanobacteria, such as the rapid-growing and nutritious *Spirulina*, have at times been proposed as a possible solution to problems of world hunger. *Spirulina* contains most of the amino acids essential to human nutrition and can be bought in dried form in packages distributed by health food stores. Just add water, stir, place in the light, and observe the sample.

In sum, photosynthetic, water-using cyanobacteria represent one of the most enduring and resilient forms of life. With a noble history stretching unbroken in the fossil record back more than two billion years, they have inhabited the planet more than 500 times as long as recognizable members of the human genus have. If you picture the history of cyanobacteria as spanning the length of a football field, the multimillion-year-long history of humans would cover only several inches of the end zone.

FIGURE 10.14
Called a hormogonium, this is a piece of filament from filamentous cyanobacterium of the family Nostocales. Hormogonia can glide away and grow by themselves. (Drawing by Ilyse Atema.)

FIGURE 10.15
Gomphosphaeria, a colonial cyanobacterium. The fluorescent glow of the bottom photograph results from chlorophyll excited by ultraviolet light that has been shined on *Gomphosphaeria*.

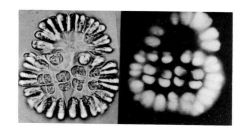

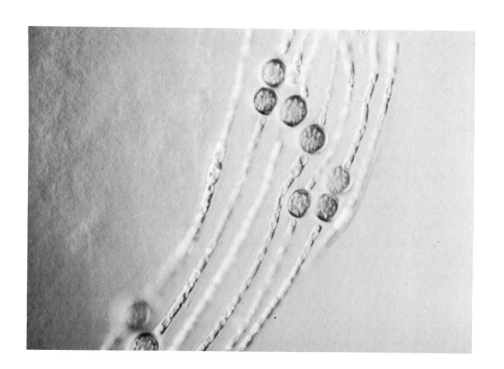

FIGURE 10.16
Anabaena, a filamentous cyanobacterium. The large rounded cells are heterocysts, specialized cells that are not photosynthetic but can fix nitrogen. (Photo courtesy of Stuart Brown.)

119

FIGURE10.17

Root nodules on a leguminous plant. You can dig up a legume such as a bean, pea, vetch, or clover plant and find them your-self. Often the nodules are pink.

■ NITROGEN FIXERS ■

Nitrogen is critical to all forms of life, largely because all the important directing molecules found in cells (DNA, RNA, and proteins) are partly composed of this element. Aside from costly manufacturing procedures, only nitrogen-fixing bacteria are capable of taking inert nitrogen from the air and making it accessible to all living organisms. Without them life would long ago have ceased because the most essential biomolecules would have broken down due to nitrogen deficiency.

Nitrogen-fixing aerobic bacteria are found all over the surface of the Earth, in both soil and water. They are especially prevalent in garden soils, on leaves, and associated with the roots of certain vegetables, called legumes. Some can survive in places where the percentage of oxygen in the air drops below the normal 21%; others inhabit oxygen-rich, highly aerated waters.

There are many groups of nitrogen-fixing aerobic bacteria, including the azotobacters, which are cyst-forming, rapidly growing, oval-shaped cells that exude profuse amounts of slime. The protective cysts are more spherical than the cells themselves. Azotobacters are sometimes associated with black pigments, which they produce and which are insoluble in water. Another kind, the *Beijerinckia*, are named for Martinus Beijerinck, a Dutch micro-biologist from Delft. In 1901 he wrote:

> *I shall call oligonitrophilic microbes those which in free competition with other microbes grow in nutrient solution where one has not willingly introduced nitrogen-containing substances but from which one has not removed the last traces of these compounds. They have the property of fixing free atmospheric nitrogen either alone or in symbiosis with other microbes.*

Beijerinck had just found such a bacterium (*Agotobacter agilis*) in water from the canal at Delft. Like *Azotobacter*, *Agotobacter* is rod-shaped and produces a sticky extracellular slime. But *Beijerinckia* cells are notable for their accumulation of internal lipid (oily) structures found at opposite poles of the cells.

The easiest way for you to find an aerobic nitrogen fixer is to dig up leguminous plants, clean the soil off of them, and look for root nodules (Figure 10.17). Nodules can be clearly seen on the roots of legumes only several weeks old. Often they are pink because of the natural pigment leghemoglobin. Cut a nodule sliver thin, and with your microscope observe the irregular-shaped "bacteroids"—swollen, nitrogen-fixing bacteria packed into root cells (Figure 10.18). You can use any of the following plants: soybeans, navy bean or any other food bean, purple vetch, clover, sweet peas, or peas. Commercially available bags of soil rich in these bacteria are available for planting legumes in pots.

Under the transmission electron microscope, aerobic nitrogen fixers can be recognized by their transformation into the large, weirdly-shaped bacteroids. They are always Gram negative and most have flagella when they live in the soil outside the nodules, before they transform into bacteroids.

The most famous form of nitrogen-fixing aerobic bacterium is certainly *Rhizobium*. It is *Rhizobium* that lives with members of the Leguminosae family, which includes such members as peas, alfalfa, soybeans, beans, and lentils. Because *Rhizobium* fixes atmospheric nitrogen, many of these important food crops are able to subsist in nitrogen-poor soils that would otherwise preclude their growth.

As you can see for yourself, in both the free-swimming bacteria form and in the swollen, captive bacteroid form, *Rhizobium* lives in the root hairs that protrude from the roots of legumes. It chemically entraps the plant into producing nodules—special *Rhizobium*-containing structures specified by bacterial DNA (see Figure 10.17). The *Rhizobium* simply remains there, holed up in the nodules. Roots decorated with an abundance of nodules, legumes provide us with the most complete plant source of nitrogen we have

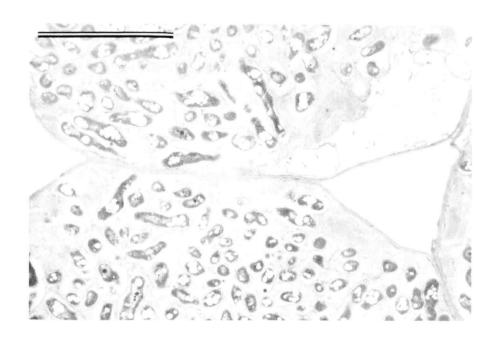

FIGURE 10.18
Bacteroids—transformed bacteria—in root cells from a nodule (like those shown in Fig. 10.17). Thin-section electron micrograph.

in human nutrition. Interestingly, *Rhizobium* living freely in the soil is *not* capable of fixing nitrogen. *Rhizobium* does have nitrogen-fixing abilities in its DNA, but these biochemical "switches" are "turned on" only by contact with plant root cells. Scientists are now working on ways to engineer the nitrogen-fixing abilities directly into non-leguminous plants, which could hypothetically revolutionize agriculture.

▪ FALSE MONADS ▪

The pseudomonads are an extraordinary mixed bag of bacterial survivors. Members of this phylum of bacteria share similar traits but probably have distinct ancestries. (*Monad* was an old name for eukaryotes, so *pseudo*monads are *false* monads or prokaryotes.)

Pseudomonads are oxygen-breathing, Gram negative rods or spheres. They are versatile microbes ubiquitous in soils and water. While the phylum encompasses a range of bacteria types, all pseudomonads are joined in their ability to respire oxygen, and most probably evolved around the time of the accumulation of oxygen in the atmosphere, some two billion years ago. However, some also can breathe without oxygen, which they replace in their metabolism with nitrate, an oxidized form of nitrogen. Pseudomonads metabolize an amazing variety of compounds. They not only can break down sugars, alcohol, and organic acids but also far more complex chemicals, such as some of those found in petroleum and the usually intractable natural fibers of plants.

The *Xanthomonas* genus of rod-shaped bacteria that form yellow colonies on agar has been found associated with and is thought to cause serious plant disease such as leaf spots. *Zoogloea* is a pseudomonad genus that demonstrates the multicellular nature of many bacteria. An individual cell starts life alone, moving in fresh water by means of a single polar flagellum. After sufficient reproduction, however, the cells come together to make flocs or globs. These flocs may be visible to the naked eye as shiny tiny scums on the surface of lakes and ponds. Cohesive structures protruding out from the center of the film help to identify the colonies as *Zoogloea*. One species of *Zoogloea*, *Z. piliperdula*, has been isolated from sunken pistons and other underwater objects in a waterworks near Berlin. Another member of the genus, *Z. ramigera*, was originally found in sewage sludge.

Bdellovibrio (Figure 6.0) are aerobic, rod-shaped pseudomonads that parasitize other, more vulnerable bacteria to survive. They reproduce inside the cells of their victims, which are destroyed when maximal numbers of *Bdellovibrio* are produced. If they attack a large spirillum, many "baby" *Bdellovibrio* come out. If their hosts are small, only a few *Bdellovibrio* emerge.

The salt-loving halobacteria also are included in the pseudomonad phylum because they are oxygen-respiring, Gram negative, rod-shaped bacteria. Nonetheless, many bacteriologists recently have begun to argue the merit of placing halobacteria with the methanogens and a few other forms into their own kingdom. This is because molecular biological studies of the RNA in the ribosomes of halobacteria indicate a close kinship with the methanogens and certain others. Most refer to this proposed new group as the archaebacteria (see pages 113–115, Methanogens). Japanese scientist H. Hori suggests such beings are better referred to as metabacteria because archaebacteria connotes "old," i.e., that they evolved first, when the molecular biological evidence simply suggests that they are different.

Whatever one chooses to call them, methanogens and halobacteria seem to have diverged quite a long time ago from the main branch of bacteria. Unlike other pseudomonads, the halobacteria not only love salt, they absolutely require it. The halobacteria inhabit salt flats, saturated salt solutions, and brines the world over. Since halobacteria produce vast amounts of warm-colored pigments called carotenoids, they are visible from airplanes and satellites as a pink film on salt flats.

Halobacters are easily collected and recognized. The best place to find them is along tidal flats where salt has precipitated. Look for color, as Charles Darwin did when he described their environment in 1832 from the voyage of the Beagle off northeastern Brazil. Put the orange, salty material under your microscope using high power. With a pipette, slowly add a drop of distilled water under the slide coverslip. Watch the salt-loving bacteria explode: they find the distilled water, containing no salt, to be suddenly lethal as a result of the change in osmotic pressure.

■ AIR-BREATHING SPORE FORMERS ■

The spore formers, or aeroendospora, are oxygen-breathing microbes that have acquired the trick of making specialized reproductive structures within their cells. These structures, called endospores, protect the bacterial progeny from death by heat and drying out (Figure 10.19).

Three genera of aerobic endospore-forming bacteria are known: *Bacillus*, *Sporosarcina*, and *Sporolactobacillus*. The bacilli are rod bacteria that pro-

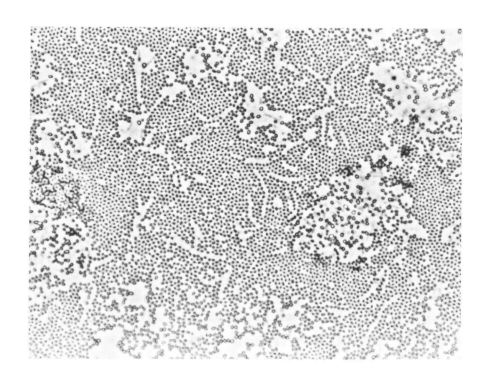

FIGURE 10.19
Stained cells and endospores of bacillus bacteria under the light microscope. When the colony is forced to dry out, the spores (round and shiny) start to form inside the growing cells.

123

duce spherical or oval endospores; each bacillus cell makes one endospore. The spore, so light it is wafted by air, may survive for years before it lands in a moist environment, upon which it immediately renews the cycle of growth. Hundreds of strains of *Bacillus* are known, making this one of the best-studied bacteria in the world.

Bacilli look like typical rod-shaped bacteria under most high-powered microscopes. Most species move by means of waving flagella. When bacilli grow in colonies they tend to form branching chains of cells; sometimes entire colonies of bacilli rotate so quickly you can watch an entire rotation in less than one minute. The mechanism and purpose of such behavior is still unknown.

The very common *Bacillus* genus of bacteria is quite important commercially. Many kinds of bacilli produce antibiotics that have been used to save millions of human lives. Furthermore, bacilli are valuable to the biosphere as a whole because they are one of the few types of organisms able to metabolize intransigent, large molecules of plant materials, such as pectin, polysaccharides, and cellulose, the main ingredient of wood. The genus *Bacillus* includes a variety of bacteria adapted to a gamut of differing salt, nutrient, and temperature conditions. For example, whereas some bacilli, called psychrophils, grow optimally at –30°C (–3.5°F), others, such as *Bacillis thermophilis*, prefer such hot locales as those found in hot springs where temperatures climb to more than 45°C (113°F).

Sporosarcina ureae is an interesting endospore-forming bacterium in that it enjoys the ability to convert urea, a constituent of urine, among other things, into ammonium carbonate. *S. ureae* grows in packets of four that may appear almost cubical under the microscope. *Sporolactobacillus*, as its name suggests, is like the fermenting bacterium *Lactobacillus*, except that it forms internal spores and is able to respite both with and without ambient oxygen. All members of this phylum can respire in the presence of oxygen; some, however, can also switch to anaerobic respiration when oxygen is not available.

Spore formers are easy to collect because of the resistance of their spores. Leave a cup of bouillon or any old soupy food exposed to the air for three days or more. Then heat to very hot (but don't boil) the mixture for 10 minutes or so. While you're at it, boil a quarter of a cup of covered bouillon for 20 minutes. Add a spoonful of the cooled soup to clean bouillon. After a few days, spore-forming bacteria will be growing in the once-clear bouillon. Put some of the broth on slides and let it dry out. Check with your microscope and you'll see the spores.

■ GRASS-GREEN BACTERIA ■

P*rochloron* and *Prochlorothrix*—the only green, or chloroxybacteria, yet discovered—may be evolutionary missing links between bacteria and plants, or rather between photosynthetic bacteria and chloroplasts, the green photosynthetic parts of plant cells. *Prochloron* (Figure 10.20) are fascinating because they combine the anatomy of a bacterium cell with the photosynthetic physiology of a plant. They perform the oxygen-producing photosynthesis of algae and plants; yet they lack nuclei, chromosomes, and mitochondria, and thus are unambiguously bacteria. In their oxygenic photo-

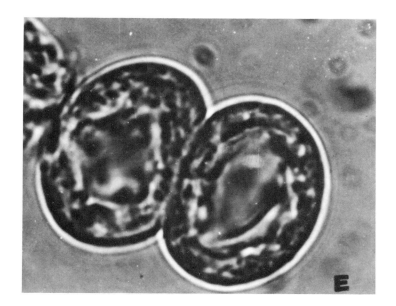

FIGURE 10.20

Prochloron, a grass-green chloroxybacterium, from the surface of a "sea squirt" or didemnid (a tunicate animal) from the South Pacific.

synthesis chloroxybacteria resemble cyanobacteria, but they contain the pigment chlorophyll *b* in addition to chlorophyll *a*. This makes them like all green algae and plants, and separates them, biochemically speaking, from all other bacteria on Earth.

Chloroxybacteria get their name from the Greek word *chloros,* meaning green. *Prochloron* has been named to suggest its possible ancestry to chloroplasts. *Prochloron* is known almost exclusively from the Gulf of California (Sea of Cortez) and the South Pacific where, indeed, like the putative ancestors to all chloroplasts, it is a symbiont. Instead of living inside larger nucleated cells, however, *Prochloron* lives symbiotically upon the surfaces and cloacae (combined genital–excretory organs) of lumpy marine animals called tunicates or, more familiarly, sea squirts.

Prochloron, and its relative *Prochlorothrix* from Lake Loos near Amsterdam, originally possessed the ability to photosynthesize with both chlorophyll *a* and *b* pigments, generating oxygen all the while. No matter the exact species of photosynthetic bacterium, the green prokaryote—then as today—could have been eaten by larger protist cells with nuclei. The smaller green cells were then "harvested" in a sort of internal agriculture. The merged progeny evolved into the green algae, as well as all plant life on Earth. If this story of the origin of green algae and plants is true, it emphasizes the advantages of forming symbiotic partnerships: *Prochloron* is now a rare bacterium, having little global impact because of its insignificant population numbers. But plant and algal life—the assumed progeny of the symbiotic merger—are as diverse and crucial as the world is wide. We eat them as vegetables and fruits, and depend on their oxygen waste to breathe. And we animals live by inhaling what may once have been mere chloroxybacterial waste, not hazardous but precious, oxygen gas.

Today, *Prochloron* may be squeezed out of sea squirts—lemon-shaped tunicates floating about the South Pacific. You must find the proper animal host first and look for a greenish tinge on it. *Lisoclinum* and *Didemnis,* organisms whose appearance and characteristics are described in marine or

125

invertebrate zoology texts, are good examples of host animals for *Prochloron*. Chloroxybacteria form a green film upon the surface of these colonial marine animals that can be scraped off and examined under a microscope. So far, however, the microbes in the film have proved extremely difficult to keep alive in the laboratory for extended periods of time. This may be due to their ancient penchant for life in close contact with others.

▪ MICROCOCCI ▪

The phylum Micrococcus derived its name from a linguistic mix: the Greek word *mikros*, meaning small, and the Latin *coccus*, which signifies berry. These berrylike bacteria often grow in clumps or packets of four, called tetrads. In addition to irregular cubical structures, some genera, such as the *Planococcus*, preferentially arrange into units of twos and threes. Micrococci are Gram positive and always require atmospheric oxygen for growth.

They are well dispersed throughout the world's surface, inhabiting both land and water regions. They are often symbiotic—and occasionally pathogenic—within larger organisms, human beings included. Although all micrococci need oxygen to grow, some can respire anaerobically as well as aerobically.

The genus for which this bacterial phylum is named is *Micrococcus* itself (Figure 10.21). Within this genus, there are several species including *M. luteus*, which may be recognized by its production of yellow pigments, and *M. roseus*, known for its reddish pigments. (Both sorts of pigments are known as carotenoids, which get their name from carrots because they give that root its orange color.) *M. radiodurans*, which superficially resembles *M. roseus*, has a cell wall with a composition unlike all other *Micrococcus* species. Some have postulated that the unique cell wall of *M. radiodurans* accounts for the ability of this bacterium to resist significant amounts of ultraviolet and gamma radiation.

Other micrococci can tolerate harsh environmental conditions. *Sarcina*, for example, not only can respire in the presence of gaseous oxygen but can ferment sugars in its absence. Still other micrococci can withstand saline environments about one and a half times as salty as seawater.

All micrococci can break down hydrogen peroxide into its nontoxic constituents, oxygen and water. If you suspect you have a micrococcus colony, you can check for their presence by adding hydrogen peroxide—obtainable in pharmacies for the treatment of cuts—to the sample and watching for bubbles of oxygen to come off. The reaction occurs just as on your skin, and for the same reason: because of the enzyme, catalase, which is present both in micrococcus colonies and on normal human skin. For details, look up catalase in one of the laboratory manuals listed on pages 93–95.

Staphylococcus is a well-known member of the micrococci phylum, important as a culprit in many human infections. *Staphylococcus* may be associated in maladies ranging from sore throats and skin patches to venereal infestations and spinal meningitis. All the same, *Staphylococcus*, like other microbes singled out and blamed as the sole cause of human ailments, is usually present and harmless on its human or animal hosts. Like other

FIGURE 10.21
Several micrococci from the genus *Micrococcus*. (Drawing by J. Steven Alexander.)

bacterial agents purportedly causative of disease, enhanced growth may be as much a symptom of a distressed immune system as its cause.

A great variety of bacteria, including "staph," are normally present in all parts of animals. Eradication of some groups may well have a beneficial effect, but it also may prove deleterious. Think of the customary tendency of some microbes to grow in the waste of others. Supposed culprits also serve to keep other potentially dangerous microbes in check. For example, although spinal meningitis is often effectively treated with antibiotics that kill bacteria, the extensive use of anti-bacterial medicines in recent decades has led to a considerable rise in the outbreak of fungal meningitis, for which there is no known effective cure.

Another beguiling member of the micrococci phylum is *Paracoccus denitrificans*. Formerly called *Micrococcus denitrificans*, this bacterium is a free-living, optional aerobe. The details of its respiratory system have been carefully studied and found to be nearly identical to that of the mitochondria of yeast and animal cells. Mitochondria, because they have their own DNA and RNA, and for other reasons, are considered to have evolved from aerobic bacteria that proliferated in the aftermath of the cyanobacterial revolution, which raised the quantity of atmospheric oxygen from nearly zero to 21%. This being the case, *Paracoccus denitrificans* so far has been the most convincing candidate for the ancestor of mitochondria. Like a staph infection, *Paracoccus* ancestors may have at first invaded or overgrown other bacteria only slightly larger than themselves. But in time *Paracoccus* found that it was better fed if it did not kill its prey but capitulated, providing the host with energy derived from the aerobic use of oxygen in return for organic acids and a place to live.

Although micrococci are common types of bacteria, you would need a strict sterile technique to grow these genera and they are very difficult to distinguish.

▪ GAS EATERS ▪

Gas eaters, or chemoautotrophs, live on air. Unlike the vast majority of beings on the planet, including all plants, animals, fungi, and protists, not a single vitamin, carbohydrate, protein, sugar, or amino acid exists on their list of recommended daily requirements. Instead, the minimum requirements of gas eaters are met by nitrogen-containing salts, oxygen, carbon dioxide, and hydrogen-rich (reduced) gases such as ammonia (NH_3), methane (CH_4), or hydrogen sulfide (H_2S). Reduced gases react with oxygen to provide a source of energy. Gas eaters tap into these inorganic energy sources and are crucial to the global recycling of nitrogen, carbon, and sulfur because they convert unusable gases and salts into organic chemicals usable by the majority of life.

Autotrophy refers to organisms that do not depend on other organisms for food, but live and grow in the absence of preformed organic compounds. Some, such as the abundant soil microbe, *Nitrobacter winogradsky* (Figure 10.22) (named after the pioneer of microbial ecology, Sergei N. Winogradsky), are even inhibited by organic compounds. In the introduction to his influential 1949 *Soil Microbiology*, Winogradsky said:

127

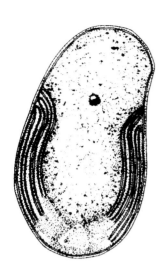

FIGURE 10.22

Nitrobacter winogradsky, a chemoautotrophic, nitrite-oxidizing bacterium. (Drawing by Ilyse Atema.)

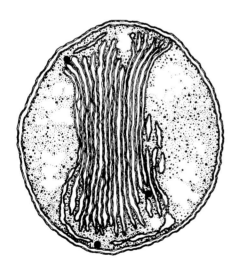

FIGURE 10.23

Nitrosococcus oceanus, a chemoautotrophic ammonia-oxidizing bacterium. (Drawing by Ilyse Atema.)

I started my work in 1885 . . . . impressed by the incomparable glitter of Pasteur's discoveries, as a young student I entered this field of investigation and I have remained faithful to it to the end. . . .

After explaining his botanical rather than medical interests in microbiology Winogradsky continues:

My first investigations dealt with filamentous bacteria found in sulfur and iron-containing springs; these were the first known autotrophs...

On Earth, there are only two kinds of organisms with self-sufficient metabolisms: the sun-loving photoautotrophs (photosynthetic organisms), which use light as their energy source, and the "rock-eating" chemoautotrophs (also called chemolithotrophs), which do not even need light but only natural inorganic chemical reactions to derive their energy. Among all the organisms on the planet that might possibly have evolved such an independent mode of nutrition, only chemoautotrophic bacteria did.

The chemoautotrophs fall into roughly three general categories. For example, those like the ones in Figures 10.22 and 10.23 live by oxidizing nitrogen compounds such as nitrite and ammonia. Those that oxidize nitrite into nitrate include *Nitrobacter, Nitrospina, Nitrocystis,* and *Nitrococcus.* Nitrobacters are rod- or pear-shaped bacteria with elaborate membranes at one end of the cell; nitrospinas are long, slender rod-shaped marine organisms with no such membranes; and nitrococci are spherical cells with dense sets of internal membranes.

The ammonia oxidizers oxidize ammonia to nitrite. These include *Nitrosomonas, Nitrosospira, Nitrosococcus,* and *Nitrosolobus.* Since they depend on the simultaneous presence of ammonia and oxygen in their vicinity, they are found at interfaces in muds near the ocean, deep in lake water just where the oxygen is deflected, and in certain soils. *Nitrosomonas* are rod- or oval-shaped; *Nitrosospira* are freshwater spiral-shaped microbes; *Nitrosococci* are spherical; and *Nitrosolobus* species are amorphic cells covered with flagella.

Chemoautotrophs that live by oxidizing such inorganic sulfur chemicals as sulfide, thiosulfate, polythionate, and sulfite are specialists that belong to entirely different groups than the nitrogen chemoautotrophs. *Thiobacterium* are rod-shaped bacteria; *Macromonas* are cylindrical with flagella ornamenting both cell ends; *Thiovellum* actually form veils—hence their name. They are elliptical cells with peritrichous flagella, meaning that they have groups of flagella scattered about their surfaces. Some sulfur oxidizers, such as *Thiospira, Thiobacillus,* and *Sulfolobus,* are spiral-shaped with polar flagella. One such sulfur-using species, *Sulfolobus acidocaldarius,* is so adapted to scalding hot sulfur springs that it dies of freezing when temperatures drop below 55°C (131°F), the temperature of extremely hot tap water.

The final category of chemoautotrophs consists of all those that grow by oxidizing methane or methanol. *Methylomonas* are Gram negative rods, while *Methylococcus* are spherical cells with a penchant for growing in pairs. By oxidizing the methane and sulfide waste that methanogens and sulfate reducers produce, chemoautotrophs help to close a dynamic cycle of global chemical interchange. The earth's surface has the trappings of a perpetually evolving machine, what with the oxidizers deriving energy from the reducers, and the reducers getting their food from the oxidizers—all ultimately driven by the beneficent luminosity of our neighborhood star, the Sun.

▪ OMNIBACTERIA ▪

Most life on Earth takes the form of some kind of omnibacteria: optionally anaerobic, Gram negative, unicellular, rod-shaped bacteria. Sporeless yet mysteriously hardy bacteria, these rods appear in nearly any water sample anywhere, if incubated under the proper conditions. Among the omnibacteria is *Escherichia coli*, the best-studied form of life on Earth. A harmless and would-be obscure dweller of our intestines, *E. coli* has become a star organism, the main source of knowledge for all of molecular biology. *E. coli* divides into two every half hour, providing immense quantities of material for study. It only requires a glass tube or flask and a sugary solution with a few salts to grow prodigiously.

If they live in animal intestines, omnibacteria are called enterobacteria. These are respiring bacteria that can switch to nitrate when oxygen is unavailable for breathing. Like the animals whose guts they often inhabit, the omnibacteria are heterotrophs: they depend upon pre-made organic compounds for both energy and growth. As the most diverse and comprehensive phylum, omnibacteria may be broadly—and, it must be admitted, arbitrarily—divided into two huge groups: the unicellular enterobacteria often associated with plant and animal diseases, and the multicellular stalked, budding, and aggregating bacteria.

E. coli is one form of the enterobacteria. Also among their ranks are *Enterobacter, Salmonella, Proteus, Citrobacter, Klebsiella, Edwardsiella, Serratia, Yersinia (Pasteurella), Erwinia*, and *Shigella*. Most of these Gram negative rods have numerous flagella. Also within the enterobacteria group are many kinds of vibrios, comma-shaped microbes with a single wriggling flagellum. Here we find *Aeromonas, Pleisiomonas, Xenorhabditis*, and *Photobacterium*, as well as *Vibrio* itself, notorious as the cause of cholera. Figure 10.24 shows a newly discovered magnetotactic vibrio from the surface sediments of the Santa Barbara Basin in California. (The black bodies inside it are actually tiny magnets.)

Vibrios are among the most exotically pigmented bacteria. One kind of marine vibrio produces bright red prodigiosin, while different *Aeromonas* produce red-yellow, orange, and purple-brown pigments, not all of which are carotenoids. Unlike chlorophylls, which function to absorb visible wavelengths and transport elections to be used for photosynthetic food production, the purpose of the various vibrio pigments is a mystery. There are even omnibacteria that are neither enteric nor vibrio but that also produce beautiful pigments. For example, aptly named *Chromobacterium* makes violacein—a bright violet, ethanol-soluble pigment.

Even more intriguing are the bioluminescent vibrios: *Xenorhabditis, Photobacterium*, and some species of *Vibrio* itself. *Xenorhabditis* shines its own light. The glowing bacterium has been found in association with microscopic animals called nematodes and with insects that feed on nematodes. Perhaps most amazing is the symbiotic lantern microbe

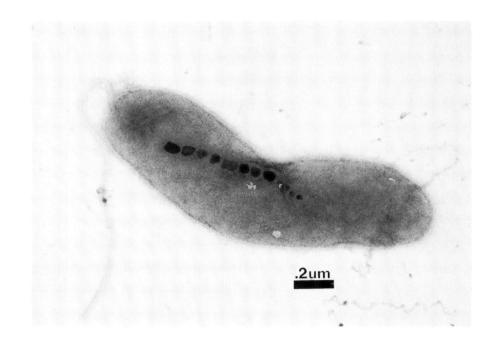

Photobacterium found shining as the lights of luminous fish. The photobacteria swarm in special pockets of the fish. The fish become luminous and can navigate in search of food at ocean depths so deep the sea is as black as space. The fish wander here as a result of this evolutionary association with the omnibacteria.

Among the multicellular omnibacteria are such forms as *Caulobacter* and *Asticcacaulis*, which extrude stalks called prosthecae that attach to rocks or other solid surfaces. Only alternate generations of prosthecate bacteria have stalks: a stalked form will divide and produce a non-stalked offspring, and so on. From budding omnibacterial cells grow protrusions, buds that swell up to become full-grown mature bacteria. Some aggregated omnibacteria are capable of depositing iron or manganese oxides around them. Others, like the flagellated cell shown in Figure 10.24, try to swim toward the north magnetic pole. In so doing they reach the mid-water oxygen interface where they can breathe in peace. Although they need oxygen from the air, they are made sick by the high quantities present at the air–water interface. How do such magnetotactic bacteria know to seek the North Pole? Somehow they manufacture inside their bodies regularly shaped little magnets composed of magnetite (iron oxide). These intracellular bodies provide us with a prime example of "biomineralization"—the formation of solid minerals by cells.

Also among omnibacteria are malignant forms such as *Neisseria*, found in genital lesions of gonorrhea patients and in the spinal fluid of people sick with meningitis. *Chlamydia*, too, the intracellular parasite that causes a very common but still rarely diagnosed venereal infection of the same name, is an omnibacterium. There are even alcohol-loving strains of omnibacteria. *Gluconobacter* and *Acetobacter*, acetic-acid bacteria that oxidize ethanol, turn sweet wine into sour vinegar.

Literally any soil, food, or water sample you look at is bound to contain omnibacteria. To distinguish them is a microbiologist's task. To provide yourself with a large sample of these bacteria, prepare an infusion. Place a few grains of wheat, rice, or wild seeds in a cup or so of some fresh water. Let it sit for a day. Cover it with clear plastic wrap and let it sit on a window sill. Then look at it, and with your microscope at high power you will glimpse the microbial world examined in such detail by von Leeuwenhoek and the other microbiologists of Delft.

▪ ACTINOBACTERIA ▪

Nature lovers know a familiar and special smell that belongs to the woodlands. Reportedly, however, this smell comes from neither leaves nor a subtle combination of flowering trees. Rather it is a product of actinobacteria and their activities within the forest floor.

The actinobacterial phylum includes the coryneforms, Gram positive straight or curved rods that sometimes form clublike swellings. Members of this phylum typically form V- or Y-shaped configurations as a result of attachment to offspring after dividing. *Cellulomonas*, which degrades the difficult-to-digest cellulose of trees, possibly effusing an aromatic woody perfume in the process, is a corynebacterium. So are the elongated, multiple-spined *Arthrobacter*, and *Propionibacterium*, the latter named for its propensity to produce propionic acid as a by-product of its sugar metabolism. (*Propionibacterium* makes acetic acid as well.)

Then there are the actinobacteria proper. Also called actinomycetes or actinomycota, these used to be considered true fungi. These bacteria mimic the growth habits of fungi—not as the result of evolutionary similarly but by assuming similar lifestyles under similar circumstances. (This phenomenon is known in evolution as convergence.) True fungi structures consist of long thin filaments called hyphae and often disseminate resistant spores, called conidia. Actinobacteria also form filaments, though they are generally thinner and shorter than those of fungi. These actinobacterial filaments consist of cells that convert into thick-walled resistant forms (unlike bacterial endospores, which form by partitioning within the parent cell). In addition, two families of the filament-forming actinobacteria actually bear their actinospores in protective structures called sporangiophores—highly suggestive of conidiophores, the spore-containing "heads" of true fungi. But since actinobacteria cells and the actinospores into which they transform have no nuclei, they are prokaryotes and thus can only be considered bacteria. Still, such spore-containing headlike structures, which disintegrate in the face of blowing winds to spread their offspring, are a solution to dispersement that has been arrived at independently and in many different organisms in the long annals of evolution.

Like other bacteria, but totally unlike fungi, some actinobacteria, even free-living ones, can fix atmospheric nitrogen into organic compounds. Actinobacteria such as *Frankia* even form bacterial nitrogen-fixing nodules with non-leguminous plants such as alder trees, especially in northern climes.

Actinobacteria that belong to a group called the Dermatophilaceae family contain species whose members are spherical and swim away to form multicellular filaments. These swimming, asexually dividing forms simulate the behavior of chytrids and other protists that evolved much later. Indeed, they are superficially so similar to certain eukaryotic cells that they have been called zoospores, a name denoting spermlike swimming cells capable of

131

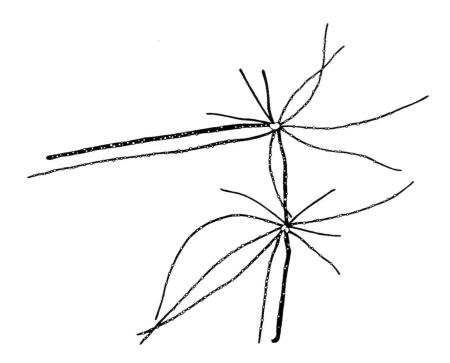

indefinite reproduction on their own that really should be reserved for protist cells. The very existence of such actinobacteria again highlights the phenomenon of convergence, by which differing organisms develop similar lifestyles under similar environmental conditions.

In spite of its fungal name, *Streptomycetes* is perhaps the most illustrious group in the actinobacterial phylum, although it cannot be cultured without a strict sterile technique. Figure 10.25 depicts *Actinoplana*, an example of actinobacteria similar to *Streptomyces*. The genus *Streptomyces* is an advanced bacterium by any standards. From the standpoint of form, it is advanced because it grows in long multicellular filaments and branches, and produces its spores in an aerial spore-bearing sporangium. From the standpoint of function, it also must be considered advanced. *Streptomyces* and its close relatives produce streptomycin and related antibiotics (kanamycin, dihydrostreptomycin) that work by chemically binding to the ribosomes of susceptible bacteria and gumming up their protein-synthetic works. Streptomycete products save millions of people from death by bacterial infection. Yet, because they have developed ways of keeping the poison from its target within themselves, members of the streptomycete group are as immune to their own toxins as we humans are, for example, when our stomach acids fail to digest our stomach walls.

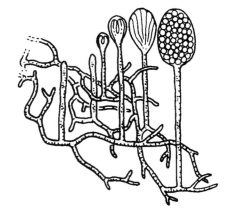

FIGURE 10.25

Actinoplana, an actinobacterium that belongs to a group whose members look so much like fungi that they used to be called actinomycetes. (Drawing by Laszlo Meszoly.)

▪ SLIME BACTERIA ▪

Multiculturity is often equated with plants and animals with their complexity and so-called "higher" natures. But this is a naive view. Some bacteria, too, are always multicellular in nature. The slime bacteria (myxobacteria) are true bacteria, and as such are considerably less complex in structure than even the simplest unicellular protists—yet they are fully capable of complex multicellularity (see Figures 5.1, 7.0).

There are two classes (sometimes considered orders) of myxobacteria: the Myxobacteriales, soil microbes that congregate to form upright bacterial "trees," called fruiting bodies—sometimes just visible to the naked eye—and the Cytophagales, gliding bacteria that do not form upright structures but that do come together into complex planar patterns. The typical myxobacterium is a Gram negative, rod-shaped aerobic cell embedded in polysaccharide slime. During times of nutrient scarcity it will glide in its sticky polysaccharide excretion and join neighboring cells of the same species. Because these bacteria lack flagella and other obvious means of motion, their gliding movement remains mysterious; they move only when cells touch each other and interact.

Myxobacteria in fruiting bodies may become myxospores—thick walled, shiny, and resistant—or they may continue growing. The fruiting bodies (actually an inappropriate, botanical expression) range from dark to brightly colored. Some fruiting bodies contain encapsulating cysts that release huge numbers of spores when wet. The spores are not carried aloft by the wind but rather migrate, gliding in colonies until harsh conditions again induce them to coalesce. Here is another striking example of convergent evolution that vividly recalls the activities of the slime molds—nucleated amebas that also congregate to form fruiting bodies in times of nutrient hardship.

Myxobacteria are among the most mysterious of prokaryotes. Because their life cycles are so complex, they are rarely seen in full in the laboratory. Many species unknown to science probably inhabit the tropics. And *Simonsiella* and *Alysiella*, two members of the family Simonsiellaceae, are multicellular filamentous gliding bacteria, chains of rod-shaped cells that inhabit the mouths of vertebrates. Although they can be grown on blood and serum in the laboratory, no one knows how they live within these animal mouths, nor what role they play there.

The Cytophagales are probably not evolutionarily closely related; but because they share the traits of gliding and heterotrophy, they are all placed within the myxobacteria phylum. *Cytophaga* is metabolically tough, able to use not only agar but hard biogenic coatings, such as chitin and cellulose, for energy. Evolutionarily related species include *Flexibacter, Herpetosiphon, Flexithrix, Saprospira,* and *Sporocytophaga.* These organisms look like blue-green bacteria that have lost their ability to photosynthesize.

The last myxobacterial family is Beggiatoaceae. The representative species *Beggiatoa* you may find growing as white slime in sulfur-rich waters and muds. They often form white rosettes, multicellular bacterial "flowers" that are light in color due to small quantities of orange carotenoid pigments. You can tell one is a member of the group by light microscopic observation alone. Look for numerous tiny sulfur globules (Figure 10.26) inside the cells.

FIGURE 10.26

Beggiatoa, a myxobacterium found in sulfur-rich waters. The light micrograph and detailed drawing show internal sulfur globules. (Drawings by Ilyse Atema.)

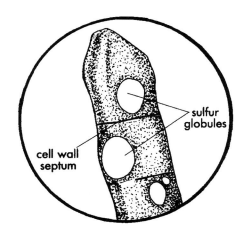

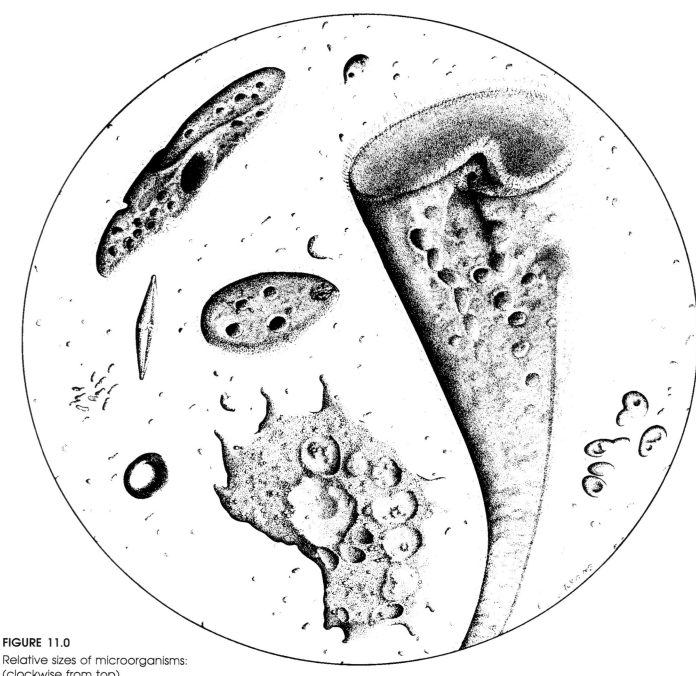

FIGURE 11.0

Relative sizes of microorganisms:
(clockwise from top)
Chlamydomonas, Stentor, yeast
(*Saccharomyces*), *Amoeba*,
fungal spore, cryptomonad, group
of bacteria, diatom, *Tetrahymena*,
Euglena, and *Paramecium*.
(Drawing by Christie Lyons.)

E·L·E·V·E·N

PROTOCTISTS

▪ INTRODUCTION TO PROTOCTISTS ▪

Between the tiny bacteria and the large, familiar organisms lies an intermediate realm of protoctists. Like plants and animals they have nucleated cells, but that is the extent of their similarity. Algae, protozoa, slime nets, slime molds—some say there are 17 different phyla of these nucleated microorganisms. (In order to grasp the range of organisms within a phylum, we should recall that we are in the same phylum, Chordata, as all fish.) Some say there are as many as 45 protist phyla. But since the science of protistology is still in its infancy (and "protoctistology" has just been conceived!), no one really knows. There are probably at least 200,000 species. The very smallest protists are bacteria-sized. And, unless they are involved in disease, these organisms, from 0.5 to about 2 microns long, are ignored by all the traditional "microbiologists"—that is, bacteriologists, protozoologists, mycologists, phycologists, and botanists. As might be expected, very few protists cause disease. Indeed, the vast majority remain undescribed.

In our guide to the world of microorganisms we can only mention the major groups of these ancestors to all familiar life forms. They vary greatly in size, shape, behavior, and life cycle (Figure 11.0). Among them are microbes that shatter the common biological assumptions that sex is necessary for variation, that all "lower organisms" are single-celled, and that different sexes always come in twos. Surprises abound for those who delve into the micocosm.

This phylum (Karyoblastea) contains only one species—*Pelomyxa palustris*, an ameba that looks like a wine flask. Amebas are single cells that have a changeable form and that feed on particles by encircling and engulfing them. *P. palustris* (Figure 11.1) is important in the study of the evolution of protists because it displays only the most basic traits of nucleated cells. Unlike the great majority of protoctists, *Pelomyxa* neither has mitochondria nor does it undergo mitosis when dividing. The organism does have a nucleus; in fact, it has many of them. This feature, combined with its large size—hundreds of times larger than bacteria—clearly make *Pelomyxa* a eukaryotic organism. (Almost 0.5 millimeters long, it looks like a dot to the naked eye.)

Despite its theoretical importance, however, *Pelomyxa palustris* has so far been found only in muddy, freshwater ponds in Illinois and a few European locations, notably Elephant Pond behind the University Museum in Oxford, England. The pond's name comes from the fact that historically it has provided a dumping ground for elephant carcasses discarded by taxidermists preparing displays at the Museum. Obscure *Pelomyxa* inhabits the top layers of the muddy bottom of the pond and has never successfully reproduced for more than a few weeks in the captivity of glass bowls on laboratory shelves.

Pelomyxa has other unique characteristics. It is microaerophilic; that is, it subsists on lower quantities of oxygen than other eukaryotes. Instead of the oxygen-respiring organelles called mitochondria, however, *Pelomyxa* has three separate types of symbiotic bacteria living in it. One of the types of bacteria is perinuclear, living in neat rings around the periphery of each nucleus. The second has a specific, recognizable structure and lives in *Pelomyxa*'s cytoplasm. Interestingly enough, these symbionts are little methane machines: like cows, *Pelomyxa* harbor methanogenic bacteria deep inside. The third sort of *Pelomyxa* bacteria is morphologically distinctive but not particularly well known. Although we do not know a great deal about these symbionts, there is some speculation that they use lactic acid, a waste product of *Pelomyxa*, for their food. If they did not use the lactic acid that accumulates in the ameba's cytoplasm, the *Pelomyxa* would be poisoned.

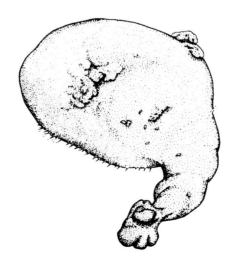

FIGURE 11.1

Pelomyxa palustris, the giant ameba, is shown from the outside. Inside it contains many nuclei and bacterial symbionts. (Drawing by Christie Lyons.)

136

Sex is unknown in *Pelomyxa palustris* and until the 1970s it was thought that the ameba had none of the waving organelles composed of ordered configurations of microtubules known as undulipodia, so common in cells with nuclei. But then, in the late 1970s microtubules and undulipodia, albeit internal, nonfunctional ones, were observed at the edges of the giant ameba. This suggests that the ancestors to *Pelomyxa* had undulipodia and that, in the evolution of protoctists, undulipodia and the membrane-wrapped nucleus preceded the evolution of mitosis, meiotic sex, and the establishment of mitochondria.

▪ SEA WHIRLERS ▪

Sea whirlers (dinomastigotes or dinoflagellates) are whirling microbes with two cell tails, or undulipodia. One of the tails is inserted in a characteristic groove of their shell, or test, which is made of plates of cellulose sometimes hardened by silica (Figure 11.2). The other whip, called the girdle undulipodium, circles the equator of the cell. *Dino,* the Greek word for *whirl* or *turn,* refers to the sea whirlers' habit of slowly turning, using their undulipodia to do so. Almost all of them are plankton—microorganisms suspended in ocean water. Dinos are a critical member of the marine food chain, providing food for larger creatures, including the largest seafarer alive, the whale.

The sea whirlers are particularly fond of warm waters where their populations occasionally grow to prodigious numbers and, along with other microorganisms, form the colorful but poisonous red tides. Some, such as *Gonyaulax,* are bioluminescent and emit a spooky greenish glow twice a day at regular intervals. In fact the timing of their luminescence persists even in locked laboratories where it has aided in the understanding of natural biological timekeeping systems, or "biological clocks." (In this regard, the sea whirlers support the endogenous as opposed to the exogenous theory—that is, that biological timekeeping exists within organisms themselves and is not subtly cued by the rotation of the Earth or other signals from the outside environment.) An organism's luminescence is a byproduct of a chemical reaction in which oxygen is consumed. It is a trait found in organisms from bacteria to protists to fireflies, fish, and some glowing mushrooms, and which probably evolved independently in many types of anaerobes as a means of ridding the cell of excessive and potentially dangerous oxygen.

Some photosynthetic dinomastigotes are a form of algae traditionally called zooxanthellae because they live in animals (*zoo*) and are yellow-brown in color (*xantho*). Very common ones such as *Gymnodinium microadriaticum* are all symbiotic within the cells of corals, clams, sea anenomes, and other soft-bodied translucent marine animals. Thus the beautiful underwater communities known as coral reefs are built largely on a foundation of energy from these sea whirlers. Some dinos also have evolved "eyes" with light-sensitive membranes. These membranes develop from the membranes of the photosynthetic plastids. While some species only have a pigmented light-sensitive layer overlain by a clear zone, others, such as *Erythropsidinium pavillardii,* have a lens and a chamber full of fluid underlain by a pigmented cup sensitive to light. The lens not only changes shape, thereby focusing, but also the whole

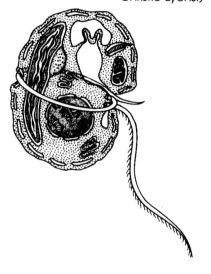

FIGURE 11.2
View inside a sea whirler (dinomastigote). Organelles shown include two undulipodia, the nucleus, a mitochondrion and a plastid. (Drawing by Christie Lyons.)

137

optical structure itself may move about the cell—perhaps on the lookout for enemies while looking for the best spots to sunbathe!

▪ AMEBAS ▪

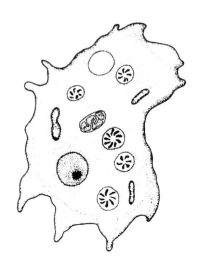

FIGURE 11.3
Amoeba proteus. Organelles shown include pseudopods, food vacuoles, and the nucleus. (Drawing by Christie Lyons.)

The amebas (Figure 11.3) are perhaps the most infamous of protoctists. Their unusual bloblike forms have served as the inspiration for the monsters of B-rated science fiction films and punk rock songs. Amebas are members of the phylum Rhizopoda, which is here defined as the phylum of food-engulfing, directly dividing protists that lack undulipodia and sexuality at all stages of their life cycle. (Throughout this book, you will note, we use the American spelling of ameba and related words.)

The amebas are distinct from the giant amebas by being slightly more sophisticated: many species contain microtubules and even chromosomes and undergo a simplified kind of mitosis. Amebas have been used extensively in studies of the relationship between nuclei and cytoplasm because they can be cut in pieces or have their nuclei removed and continue to live. Indeed, it is possible by microsurgery to remove a nucleus from one ameba and place it in another with nearly no risk of death. While some amebas are naked, others are enveloped in tiny shells, called tests, craftily formed from sand and the carbonate shells of other organisms cemented together with organic glue. Some microfossils over a billion years old—amorphous spheres—have been interpreted to be the tests of shelled amebas.

Amebas are identified by their false feet or "pseudopods"—pointy, wiggling protrusions that emerge from the cytoplasm and help them move about and feed. For example, *Amoeba proteus*, found on vegetation decaying at the bottom of freshwater streams and ponds, is a naked polypodial ameba with many moving, flowing, feeding pseudopods. From exterior to interior, the jelly-like cytoplasm of an ameba is divided into a universally thin plasma membrane; a clear, stiff layer called the ectoplasm; and a central, granulated endoplasm containing the nucleus. Tiny fibers made of the actin type of protein that is part of animal muscle tissue are found in the clear zone when it is examined with the electron microscope. Take a sample of rotting vegetation from a pond or stream and you'll be likely to observe bloblike "giants" of the microcosm. Amebas are unmistakable as they move forward and back slowly using pseudopods.

The ameba's ingestion and excretory systems are simple. The ameba has no mouth or anus but surrounds food particles with its wriggling cytoplasm, forming a sac, or vacuole, around the ingested material. Enzymes are injected into the new vacuole and the food is directly absorbed across the vacuole membrane for distribution to the rest of the cell. Waste products are diffused from many points in the cell directly into the surrounding water.

Under certain environmental conditions, many kinds of amebas can undergo encystment. They become spherical, lose much of their water, and secrete a new hard wall—an organic protective covering known as a cyst. Using this tactic, the species *Eatamoeba histolytica* can cause amebic dysentery—an abdominal sickness characterized by sharp pains when the microbial culprits, resisting digestion by cyst production, germinate inside the

138

guts of their animal hosts. *E. histolytica*, some strains of which have no mitochondria or microtubules, is one of six species of amebas normally found in the human alimentary tract.

Some amazing discoveries concerning symbiosis have been made in the study of amebas. For example, infected *Amoeba proteus* have evolved in the laboratory to become a new species of amebas dependent on the very bacteria that caused their infection. And all *Paramoeba eilhardi*, a marine microbe, contain two large intracellular packages of an obscure nature that nestle next to the nucleus and stretch and divide when the nucleus divides. These probably are ancient prokaryotic symbionts. *Paramoeba eilhardi*, in addition, is sometimes killed by certain marine bacteria that are only able to reproduce in its nucleus. Present in sea water, these bacteria, like other bacterial food, are eaten by unsuspecting amebas. The bacteria penetrate the membranous walls of an ameba's food vacuoles and crawl through the cytoplasm to the nucleus where they reproduce in such huge numbers that the ameba bursts open, killing itself and releasing them.

▪ GOLDEN ALGAE ▪

G olden algae (Figure 11.4) are beautiful—but you'll need a specialist to identify them for you. Although there are more than 300 species, all golden algae can be distinguished by their golden-yellow plastids, called chrysoplasts, that contain certain yellowish-brown pigments, including chlorphyll *c* and fucoxanthin. In addition, all of them use fat in the form of oil droplets as a food reserve. Most golden algae, or chrysophytes (Greek for "yellow plants"), are freshwater organisms, although a few silica-scavenging marine species are known.

Although many golden algae are single-celled, reproducing directly without benefit or need of a sex life, they do tend to form multicellular collectives. The colonies reproduce in two standard ways. In the first, a single swarmer, or heterokont, cell—a cell swimming by the use of two waving undulipodia—will break apart from the multicellular assemblage and then, after losing its undulipodia, divide to form new offspring algae. The second type of reproduction simply consists of the passive breaking off of a cluster of connected cells to form a new colony. Environmental disturbances, such as wave action, can aid in this reproductive mode.

Several large groups of golden-yellow algae are recognized and distinguished by their form. The Chrysomonadales contains mostly single-celled forms and their multicellular cousins, which have evolved both loosely branched and compact spherical configurations. Members of Chrysosphaerales are often spheres that release heterokont cells, which leave their parent colony to start new spheres. The Chrysotrichales are filamentous golden algae. Some form long tangled threads, while others grow in multiple branches.

The golden algae that live in the ocean, called silicoflagellates, take in sea water and scavenge silica from it. Although silica is the most common mineral in the Earth's crusts and is the mineral constituent of sand, it does not dissolve easily in water. Golden algae are able to precipitate silica particles out of solution to construct amazingly detailed and elaborate surface shells. Indeed, along with such other protoctists as actinopods and diatoms, and such animals as glass sponges, these algae keep the levels of soluble silica in the

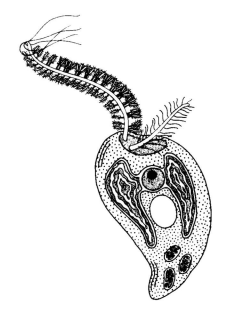

FIGURE 11.4
Golden algal heterokont swimmer. Flimmers, plastids, the vacuole, and the nucleus are shown. (Drawing by Christie Lyons.)

world's oceans so low as to be undetectable by chemical means. The microscopic shells of golden algae are so distinctive that ancient ones—over 500 million years old—have been found in the fossil record. Here, as with the coccoliths and magnet-making bacteria, we have another striking example of biomineralization, the mysterious fabrication of hard, distinctive minerals within and by cells. Exactly how the algae make such intricate silica coatings inside their cells is a mystery made more impenetrable by the extreme difficulty of growing most golden microbes in captivity.

Microscopic freshwater golden algae produce hard shells in order to survive winter and periods of dryness. These are known as statocysts and are coated with silica and iron minerals. Some statocysts have been preserved as fossils while others have been tentatively traced to the ancient spherical bodies of various sizes found in rocks. These also are thought to be fossil forms, but without modern analogues in some cases.

▪ CHALK MAKERS ▪

Chalk makers (haptomonads, haptoprotists, or haptophytes) are the world's primary compilers of calcium carbonate, otherwise known as chalk or limestone, the primary material of sea shells. In large part, they made the White Cliffs of Dover on the east coast of Great Britain. Indeed, these tiny chalk makers are so industrious in certain places that blooms of them in the ocean have been clearly detected from space by orbiting satellites.

Only recently have scientists unraveled the two distinct phases in the life cycles of chalk makers. On the one hand, micropaleontologists (those who study microfossils) were well aware of highly ornate spheres and radial forms made of chalk, clearly made by living organisms, which they called coccoliths (Figure 11.5). They even knew of coccolithophorids, the photosynthetic protists that made these structures. Cell biologists and algologists, on the other hand, had studied haptomonads, relatively fragile, golden algae-like microbes. In the 1970s, it was discovered that the two forms were part of the same being. The beautiful coccolithophorids were the durable resting stages, the "houses," of free-swimming or attached haptomonads.

Similar to golden algae, chalk microbes in their swimming stage have yellow plastids and two backward-directed cell tails (Figure 11.6), but their cell structure, which contains sacs and fibers for making the chalk pieces, is unique. They also have haptonemes—threadlike, often coiled appendages at their front ends used by some as anchors to attach to rocks or other stable objects. When a chalk maker transforms from a free swimming haptomonad into a coccolithophorid, it produces tiny slats of chalk called plates. The pattern of plates (Figure 11.7) is a feature of identification for these protists because different species and genera produce different patterns. In fact, chalk makers are so abundant and their skeletons so specific that geologists have used them to mark fossil time.

While the purpose of the intricate symmetrical displays of chalk slats is unknown, with them chalk makers are capable of withstanding conditions far harsher than they can in the free-swimming haptophyte stage. One hypothesis suggests that the ornate structures may work as miniature venetian blinds, adjusting the amount of sunlight that reaches the light-sensitive plastids.

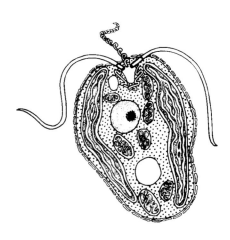

FIGURE 11.6

Golden-yellow haptomonad, swimming stage of coccolithophorid (chalk maker). Notice how the coccoliths (scales) line the exterior surface of the protist. (Drawing by Christie Lyons.)

140

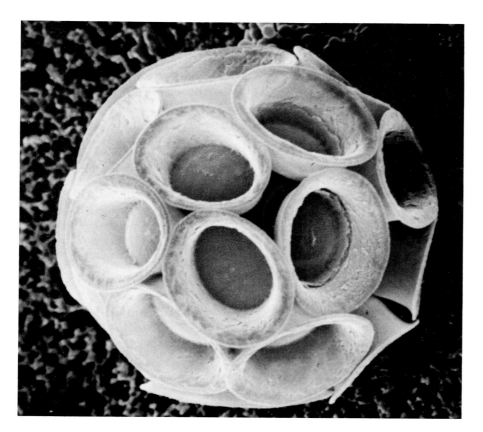

Because all the light-sensitive conversion of gas and water to food (photosynthesis) goes on in the plastid, the amount of light impinging on the plastid is crucial to the functioning of the cell. Light control is imporant to light-sensitive membranes, as those of us with eyes and sunglasses know very well. The idea that coccoliths use their slats as microscopic venetian blinds may even be correct.

▪ EUGLENOIDS ▪

Euglenoids are organisms like *Euglena*: elongated cells with a single nucleus, grass-green chloroplasts, and an eyespot (Figure 11.8). Euglenoids are easy to recognize because they are completely green, have a single undulipodium at the front end, and a flexible body that easily changes shape. They also have an internal undulipodium cell whip segment that may, by transmitting light signals from the nearby eyespot to other parts of the cell, play a role in their behavior. Euglenoids, which always reproduce by dividing along the length of the cell, are incapable of sexual encounters.

Because most euglenoids photosynthesize, they traditionally have been considered tiny green plants by botanists, members of the algal division Euglenophyta. Meanwhile, because they exhibit directed movement and animal-style ingestion, zoologists have professed *Euglena* to be single-celled animals, belonging to the protozoan group Euglenozoa or Euglenida. Really, of course, such organisms are neither animal nor plant; they are better considered members of their own line in the protoctist kingdom. Euglenoids are not ancestors to either plants or animals. Rather, they share a common ancestry as descendants of an ancient lineage of protists, or cells with nuclei.

Although, like green algae and plants, the euglenoids contain chlorophylls *a* and *b*, they also contain such esoteric pigments as diadinaxanthin and diatoxanthin, as a result of which tremendous populations of some euglenoids can be seen as reddish blooms in lakes or ponds, usually but not always during the summer. These unique beings also differ from green algae and plants because they lack rigid cell walls composed of cellulose—the tough carbohydrate that combines with lignin to give sturdiness to wood. Instead of cellulose, euglenoids have protein coverings, called pellicles, on their surface in addition to their outer membranes. These serve to increase the cell's plasticity. The pellicle is thought to be composed of rigid yet moveable circumferential plates. Finally, unlike algae and plants, euglenoids do not store starch but keep their food readily available in the form of a complex carbohydrate called paramylum.

Some euglenoids have proved to be ideal laboratory microbes for the study of chloroplasts and their relation to the rest of the cell. The most famous euglenoid in captivity, *Euglena gracilis*, can tolerate exposure to high temperatures, ultraviolet light, chemicals, or darkness. It can tolerate acid, neutral, and alkali solutions. A weed of the vast microbial garden, *E. gracilis* is found in virtually every puddle, pond, and lake, usually mixed in with the greenish covering. Collect some pond scum and look for the organism in its swimming phase after you have kept the sample for a few weeks on a sunny windowsill.

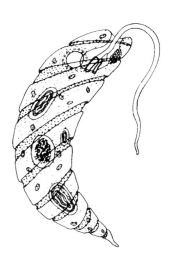

FIGURE 11.8
Euglena gracilis. The pellicle, two undulipodia (only one sticks out of the cell), nucleus, mitochondrion, and two chloroplasts are shown. (Drawing by Christie Lyons.)

142

Under certain conditions *Euglena gracilis* "turns off" or even permanently loses its chloroplasts. If exposed to the dark for prolonged periods of time, for example, this euglena turns from a "plant" to an "animal," according to the old classification system: photosynthesis halts after several cell divisions and the green chloroplasts disappear; the euglenoid then either lives by ingesting particulate food or by the uptake of dissolved food molecules. The process is reversible, however. If re-illuminated, the blanched chloroplasts become visible, turning pale green within hours. After a couple of days in the light the bright green color is restored. The process is not unconditionally reversible, however. If irradiated briefly with the ultraviolet light of a germicidal lamp, this same *Euglena gracilis* will lose its chloroplasts permanently.

▪ CRYPTERS ▪

Abundant in the world, crypters (cryptomonads) are tiny moving protists known for their crypts, or oral grooves, used in feeding (Figure 11.9). As with other protists, the classification of crypters has been wrenched between the zoologist, who calls them protozoans of the order Cryptomonadida, and the botanist, who claims them as unicellular plants of the class Cryptophyceae. Add to this the fact that some of these prevalent beings live as algae in the world's oceans—the province of marine biologists—and others live as parasites within domesticated animals—the domain of parasitologists—and you have what appears to be a somewhat schizoid organism to study and classify. Wherever they belong taxonomically, however, these beings are found in moist places all over the Earth.

The cell structure of crypters is unique. Two undulating tails protrude from the base of the crypt, propelling it along. Even in photosynthetic species, the crypt may be full of trichocysts—dartlike objects that shoot toxins—and the bacterial prey that has been poisoned by them. Although they have plastids, many photosynthetic crypters supplement their diet by absorbing food from the surrounding medium. Indeed, the presence of both purple or red (phycocyanin-containing) and green or yellow-green plastids suggests that the ancestors of crypters ingested various types of photosynthetic bacteria. Alliances between eating and eaten organisms are common in evolution. One suspects that crypters fed on various sorts of bacteria that were ingested but that developed a resistance to being digested. Live food that reproduced in captivity accounts not only for the various sorts of pigments found in cryptomonad plastids but also for the simultaneous presence and use of feeding and photosynthetic organelles. The crypters, like the dinomastigotes, euglenoids, and many other protists, are neither plants nor animals. Rather, they are former carnivores, tending and feeding their intracellular gardens, which in return provide them with energy and food.

What distinguishes crypters from all other life forms is their peculiar mode of cell division. Immediately prior to reproduction, two new undulating tails with a new crypt appear next to the old undulipodia and crypt. Then the crypt does a funny thing: it rotates into an upside down position and slowly migrates to the opposite end of the cell. Small knob-shaped chromosomes form from chromatin in the nucleus and make their way toward opposite poles

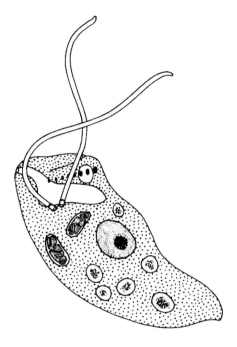

FIGURE 11.9
Cryptomonad. Shown are the crypt, undulipodia, nucleus, mitochondria, and vacuoles. (Drawing by Christie Lyons.)

143

of the nucleus. The nucleus divides first, followed by the rest of the cell. The offspring cryptomonad typically forms a plane of mirror symmetry with its parent. A common ancestry of crypters is apparent by the fact that they all share this unusual mode of asexual reproduction.

The easiest cryptomonad for you to find is one without plastids (though it may someday acquire them!) called *Chilomonas paramecium*. Let some damp plant material putrify for a couple of weeks, keeping it wet. Look for many swimming cells up to 40 microns long with two undulipodia and crypts. Other photosynthetic cryptomonads sometimes form purple and bluish blooms along sandy beaches. The best way to find these is to take home a jar of colored sand, keep it on a sunny windowsill, and prevent it from drying out by occasionally watering it with sea water from the site at which you collected the sand.

▪ MASTIGOTES ▪

In the microbial world there are neither plants nor animals—only bacteria and thousands of kinds of nucleated microbes whose true nature is complex and still under intense investigation. The mastigotes are a prime example of the confusion surrounding the eukaryotic microorganisms. An eclectic group of unicellular organisms bearing from one to thousands of cell whips or undulipodia, these rapid swimmers, or undulators, would be considered algae if they had eaten and then symbiotically cultivated blue- or grass-green photosynthetic bacteria. Because they lack color and photosynthesis, however, they often are thought of as little animals and are called zooflagellates. One of the root words of this apellation, *flagellum*, is inappropriate because rotating rods composed of flagellin proteins are structures found only in bacteria. But *zoo* is just as bad: it connotes *animal*, yet none of these protists develops from the fusion of egg and sperm into a blastular embryo, the defining trait of animalness. By contrast, the term mastigote, which comes from *mastix*, meaning whip, is more appropriate. Mastigotes are undulipodiated cells (Figure 11.10, 2.3) and no flagellin has been found in their whipping organelles. While professional biologists may be bound to tradition, amateurs should take heed not to be burdened or misled by entrenched terms.

Mastigotes, then, are nonphotosynthetic, undulipodiated protists. They form a huge group of microbes, some of which are of special interest to medicine. The major groups of mastigotes are amebomastigotes, diplomonads, retortomonads, kinetoplastids, bicoecids, opalinids, choanomastigotes, pyrosonymphids, and parabasalids. To amateur naturalists using a light microscope, whiplashing mastigotes look very much alike. But the electron micrographs taken with an electron microscope reveal a great deal of diversity in the fine structures of these organisms.

Diplomonads usually have two nuclei, which look like eyes, and eight undulipodia. They live in the digestive tracts of insects, rats, humans, and other animals. In turkeys, the diplomonad *Hexamita meleagridis* is the purported cause of an infectious and fatal form of enteritis. *Giardia* is the most notorious diplomonad. A bane to tourists, especially in eastern Europe, it causes severe endemic diarrhea.

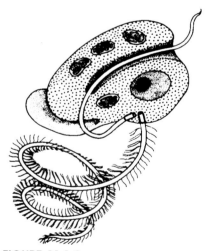

FIGURE 11.1 0
Bicoecid rnastigote. Shown are the flimmers, on one of the undulipo-dia, the nucleus, and four mito-chondria. (Drawing by Christie Lyons)

Amebomastigotes (ameboflagellates) are free-living freshwater or parasitic microbes. They are also masters of transformation. If there is no food in their vicinity, they grow cell whips and swim away to hunt for microscopic morsels. As soon as they find their bacterial food they quickly retract their undulipodia and become amebas again. Many mastigotes can exist for some time without food. In just ten minutes, they can transform their entire bodies into desiccation-resistant cysts, which permit them to wait until food and water come their way.

Other mastigotes, the opalinids, are parasitic in the rectums of fish, reptiles, and especially in such tailless amphibians as frogs and toads. Opalinids are mouthless, flattened microbes that absorb their food directly through their surfaces. They swim in a spinning manner, and it is common for nuclei in them to divide without the rest of the cell doing so. To find them, pretend you are van Leeuwenhoek and take a look at the rectal fluid of a frog.

Free-living choanomastigote (choanoflagellate) cells bear a striking resemblance to the individual cells of sponges. Sponges, one of the simplest kind of animals, are considered to be evolutionarily divergent from all other animals. Like individual sponge cells, these mastigotes have a hard structure known as a lorica at the base of a smooth-tapered undulipodium, which can be pulled into the cell. Many choanomastigotes stand on cell stalks called peduncles, and some form colonies. It thus seems probable that sponges, such as those fished from the bottom of the Mediterranean to be sold to people to absorb liquids spilled over kitchen countertops, descended from choanomastigotes.

Kinetoplastids are the best known of all the mastigotes. These asexual protists contain a special large mitochondrion, called the kinetoplast, which interacts with the undulipodium and its kinetosome, or basal body. You can find bodos, one of the few free-living types of kinetoplastid, swimming in stagnant waters. You might even find *Cephalothamnion cyclopum*, a multicellular organism that appears to be many bodos attached in unison to the body of *Cyclops*, a famous one-eyed planktonic crustacean named after the unioptical monster of Greek mythology. *Cyclops* is found in streams and brooks, and the bodo-like organisms live attached together at the summit of a common stalk.

But the most studied kinetoplastids are parasites associated with the horrible tropical diseases leishmaniasis, sleeping sickness, and Chagas' disease. Some kinetoplastids have evolved elaborate life cycles to insure their existence as they infect, debilitate, and sometimes kill a sequence of animal hosts. Sexually reproducing, parasitic trypanosomes, for instance, are capable of developmental changes in response to the conditions encountered in their different hosts. The kinetoplast of *Trypanosoma*, for example, becomes diminished in size as the parasite passes from insect salivary gland to the mammalian bloodstream and back again to insect salivary gland when next the mammal is bitten. Such protoctists demonstrate the fatuity of thinking that evolution has stopped with humans: the protoctists march relentlessly on, trying out new life cycles and forms of cell organization. Most of their sex lives and their developmental patterns are very different from ours.

By far the best way for you to see mastigotes is to find some dry wood termites (*Kalotermes*, *Incisitermes*, *Coptotermes*) or subterranean termites

(*Reticulitermes*). Keep your termites in flat glass refrigerator boxes in their original wood. Don't ever let the box dry out but don't water it too frequently or you will have fungi, not termites. It is best to keep water in small open glass containers within the box so that the air inside the box remains moist but the wood itself doesn't soak.

When you are ready for mastigote study you will need two pairs of tweezers. With one, hold the head of the termite. With the second, pull at the white material protruding from the anus until the entire digestive system comes out as a long white tube that is fat in some places and skinny in others. Puncture this tube at its fattest part and gently remove the fluid that comes out with a small dropper. If more moisture is needed, use a drop or two of your saliva to moisten the preparation. Seal the edges of your coverslip with a thin layer of vaseline. This keeps unwanted oxygen out for a while. Observing this preparation over time should allow you to distinguish the pyrsonymphids, with their bending "axostyles" running the length of their bodies, from the myriads of parabasalids, some of which, like trichonymphids, are covered with thousands of undulipodia. You may even be able to see particles of wood through the transparent bodies of *Trichonympha* (see Figure 2.0).

▪ YELLOW-GREEN ALGAE ▪

What yellow-green algae (Figure 11.11) love best are the muddy waters and drying muck of ponds, streams, lakes, and rivers. Try to find them along the banks of ditches and freshwater swamps. About 6000 species of yellow-green algae, or xanthophytes, are known. Once classified as green algae due to structural similarities with chlorophytes, yellow-greens are known to possess distinct cell organization. Their plastids are called xanthoplastids, from the Greek *xanthos*, meaning yellow. In another example of symbiosis, yellow-green algae combine heritages: their plastids seem directly related to the yellow-green plastids of the eyespot algae whereas the remainder of their cells more closely resembles golden algae.

Like golden algae, yellow-green algae have a heterokont structure, which is to say their propagules are propelled through water by two waving cell tails. One of the undulating tails is longer than the other and is elaborated with tiny lateral hairs called flimmer, or mastigonemes. That yellow-green algae share this distinctive heterokont design with select members of other lineages (see Figure 11.5; also Egg Molds and Parasitic Slime Molds) indicates, in all likelihood, a common origin. Yet the yellow xanthoplast portion of the cell represents a unique addition. Like all cells with nuclei, yellow-green algae are living examples of combinatory play, of evolution proceeding by mergers and acquisitions of living and semi-living parts.

Like the golden algae, yellow-green algae store food as oil droplets not starch as plants do. Also, like their golden cousins, yellow-greens can encyst. They possess the ability to survive the long hard winters or other detrimental conditions by forming protective cysts encrusted with minerals of silica or iron.

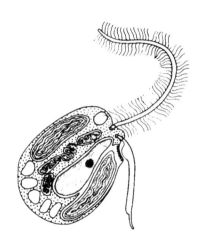

FIGURE 11.11

Xanthomonad, a yellow-green alga. Organelles shown include the undulipodia, nucleus, mitochondria, and two chrysoplasts. (Drawing by Christie Lyons.)

146

Yellow-green algae love fresh water and range in appearance from mildly gregarious swimming cells to sedentary multicellular colonies. Apparently asexual, they display quite a repertoire of structural differentiation, but they haven't been caught in a sexual act.

You can look for *Botryococcus* by collecting the greenish scum at the edge of ponds. *Heterodendron* combines filamentous cells to build an intricate, albeit horizontal, tree. *Botrydiopsis* is composed of spherical cells growing like a bunch of grapes. Perhaps most impressive is *Botrydium*. In the unlikely niche of drying muds, *Botrydium* has an upright and wrinkly but rounded appearance—somewhat like a limp balloon. By reproducing, this strange structure may grow to nearly a meter in length. Branches called rhizoids house hardened cysts and wisp down from the base of the "balloon." When the resistant cysts are wetted, they transform into heterokont cells. Looking for the moment much like other dual-whipped protists, they then swim away to form new multicellular structures in their own bizarre likeness.

■ EYESPOT ALGAE ■

Not much is known about the natural history of this recently discovered group of algae. Without excellent microscopy you won't be able to distinguish these beings from other mastigote algae in freshwater samples. Technically known as Eustigmatophyta, or eustigs for short, the eyespot algae are prolific in both fresh and marine waters as plankton, meaning they float on the surface of the sea. In the ocean, they form the basic foodstuff of many larger organisms. Some eyespot algae contain silica deposits and no sexuality has ever been reported in them.

Eyespot algae get their yellow-green color from xanthoplastids, like yellow algae. Because they share this type of plastid, eyespot algae and yellow-greens traditionally have been lumped together in a single algal division. Spectroscopic methods have revealed that the pigments of eyespot algae include chlorophylls a, c_1, c_2, and e, and beta carotene, the same as in yellow-green algae. Various species may contain other carotenoids and miscellaneous xanthin pigments such as violaxanthin, epoxanthin, diadinaoxanthin, and diatoxanthins—also very much like yellow-green algae.

Where the eustigs differ from the yellow-greens is in their eyespots and other details of cellular organization (Figure 11.12). Electron microscopy has located a concentration of photosensitive carotenoid-rich cells called an eyespot near the posterior end of these protists. In contrast to the long and short undulipodia of the heterokont xanthophytes, eyespot algae usually have only one undulipodium and a vestigial undulipodial swelling. The undulipodial swelling may play a role in a photolocation system. Through it, the eyespot may sense and transmit messages to the undulipodium so that the microbe can navigate toward the light in lakes and ponds. Sometimes eyespot algae have a second, smooth-surfaced undulipodium. But this second undulipodium is not associated with an undulipodial swelling, and so is probably not directly sensitive to light-related signals coming from the eyespot.

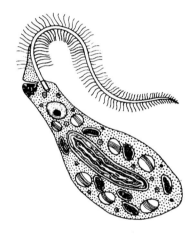

FIGURE 11.12
Eyespot alga, a eustigmatophyte. The eyespot, a plastid, seven mitochondria and an undulipodium are shown. (Drawing by Christie Lyons.)

147

Of all the free-living cells with nuclei, none are more enchanting than diatoms (Figures 11.13-11.15). These architectural masters are unrivaled in beauty and form. With cell walls, or frustules, hardened by silica—the main substance of opal and ruby, as well as of glass, petrified wood, and sand—diatoms seem to combine the symmetry of a crystal with the delicacy of a flower. Some look like sand-dollars, others like royal crowns. Indeed, natural craftsmanship has rendered the markings of diatoms so detailed and exquisite that they are often used to test the efficacy and focusing powers of new microscopes. Prolifically reproducing outside the visible realm, diatoms are ubiquitous in the world's oceans. So numerous are they, in fact, that diatomaceous earth, composed of their fossil remains, has left mineral beds over 1000 feet thick, such as in Santa Barbara County in southern California. Diatomaceous earth is used in a large number of commercial products and processes such as paints, varnishes, filters, and mild abrasives like toothpaste and metal polish. At high temperatures it is a more effective insulator than either asbestos or magnesia.

While here we consider diatoms as protoctists, traditionally they are classified as plants within the algal division Bacillariophyta. An estimated 10,000 species of them are known. Most diatoms are marine algae. They are so effective at sponging dissolved silica out of sea water to make their tiny intricate skeletons that once they have grown in a sample of water no silica can be detected remaining, even by sophisticated chemical instruments. Many kinds of diatoms live in fresh water, too.

All diatoms fall into two great classes, both photosynthetic: the centric and the pennate diatoms. Centric diatoms are radially symmetrical while pennate diatoms have bilateral symmetry and are often shaped more or less like boats or needles. The diatom frustrule is composed of two sections, called valves. Pennate diatoms, such as the colonial form Bacillaria, have a slit between their valves, called a raphe, along which they move. By extending themselves along their raphes, colonial diatoms can reach out from rocks on the shore into more rapidly moving water in order to obtain the phosphorus, silica, and other mineral nutrients they require for growth.

The diatom valves separate in reproduction, one valve going to each offspring, after which each divided cell regrows a smaller valve to fit into the original one. Succeeding generations of diatoms are therefore smaller than their parents. Such a diminishing pattern of reproduction could not have been sustained throughout evolution, so pennate diatoms, after shrinking to almost two thirds of their size in successive cell divisions, then reproduce sexually. Each diatom body converts into a naked cell capable of fusion with a mate. The fusion allows them to regain their former size by producing a large, naked, valveless diatom called an auxospore. Centric diatoms can form auxospores without undergoing meiotic sex or fertilization.

Usually tan or brown in color, diatoms previously were grouped with the golden algae because they contained golden-yellow pigments and, like golden algae, they stored their photosynthetic food in the form of the oil chrysolaminarin. Yet microscopes and further research have opened windows to the subvisible world and it now seems clear these microbes belong in a phylum of their own.

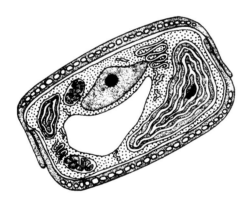

FIGURE 11.1 5

Centric diatom showing pillbox arrangement of silica frustules (shells or valves). Organelles shown include the single nucleus, two plastids, central vacuole, and two mitochondria (Drawing by Christie Lyons.)

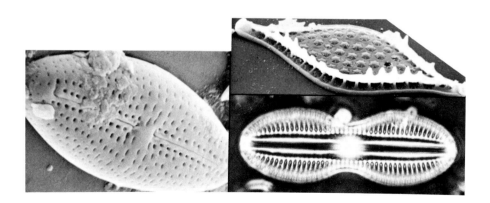

FIGURES 11.13 / 11.14
The extraordinary diversity of diatoms. (All micrographs courtesy of Prof. Frank Round except bottom right which is courtesy of Stjepko Golubic.)

To find diatoms, look for any coating that is brownish, yellowish, and crusty in shallow sea water, especially near the surface or sticking out at high tide. It is bound to contain diatoms. The coating can be on large algae, clams, or rocks. Collect the coating and its substrate (whatever it is growing on) in a clean bowl of sea water and bring it home. Drain off the excess water and in the shallow layer of water left float a few clean glass, not plastic, slide coverslips. The diatoms will actively attach to the clean glass and coat it. In a day or so take up the coverslips and use them to prepare slides. Sand grains washed with clean sea water also often have diatoms attached to them. If you see a brown crusty coating on an animal shell or skin, such as a mussel or oyster shell or fish scale, scrape it off with a scalpel. Spread a thin layer of this on a slide and observe. With luck, you will see what have been called "jewels of the sea." The diatoms will continue to live in the collection bowls if the bowls are left in indirect sunlight and continuously dampened with clean sea water. (Note: these instructions are for marine diatoms but the strategy for lake and pond diatoms is the same. The only difference is that lake or distilled water should be used rather than sea water.)

▪ BROWN SEAWEEDS ▪

The brown seaweeds, multicellular giants, really are a form of algae. Also called brown algae or phaeophytes, they are the largest members of the protoctist kingdom. Their color, ranging from olive green to sienna, depends upon the proportion of green chlorophyll *a* to brown fucoxanthin pigments. (Brown seaweeds contain chlorophyll *a*, but they lack the chlorophyll *b* present in all green algae and plants.)

Although commonly found washed up on shore, the natural habitat of brown seaweeds is the open ocean where bladders of carbon monoxide keep them afloat. Brown seaweed of the genus *Sargassum* float in a tremendous assemblage far out in the Sargasso Sea. While some brown seaweeds are small and thrive on the surface of plants, giant kelps such as *Macrocystis*, *Laminaria*, and *Nereocystis* grow several hundred meters in length, larger than an American football field. Rockweed, still another kind of brown algae, grows attached to the crags and rocks of temperate coasts.

Brown seaweeds are almost true plants, but not quite because they do not develop from embryos supported by maternal tissue. However, their reproduction tends to be sexual. Eggs, spherical or oblong in shape, and sperm with two undulating tails, one of them adorned with the lateral fuzz known as flimmer, are formed. This heterokont structure of brown seaweed sperm suggests a shared ancestry with the golden and yellow-green algae.

Once a brown seaweed egg is fertilized, the alga begins to grow into a mature plantlike form consisting of leaflike blades and structures reminiscent of stems and roots. At this stage in the life cycle, the brown seaweed may grow very large. Special reproductive structures, called sporangia, grow on its thallus, or body. The thallus looks like a gigantic leaf, but lacks the veins of conductive tissue found in plants and so cannot be considered a true leaf. The sporangia produce special mastigote cells, or zoospores, that look like sperm but grow directly, without fertilization, into multicellular organisms. In some brown seaweeds the beings generated by zoospores closely resemble those

150

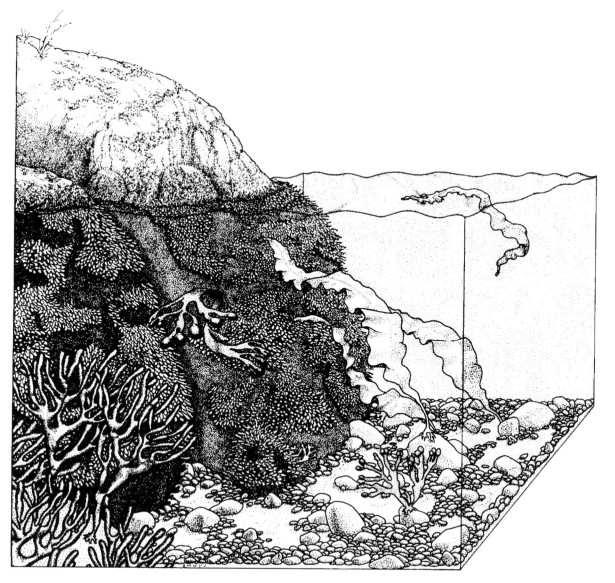

FIGURE 11.1 6

Common algae of the littoral zone along Atlantic coast (cutaway view). The long alga floating in the water and anchored to underwater rocks is the brown alga *Laminaria*. Shown clinging to the vertical wall of rock is *Fucus*. *Codium*, a branching green alga, is shown in the bottom left corner. The red alga *Gracilaria* is shown in the bottom right. (Drawing by Christie Lyons.)

grown from fertilized eggs. In others, the differences between these two phases of the life cycle are so great that for years they were not even recognized as the same organism.

The commercial importance of these protoctists is substantial and may become more so. Brown algae provide a source of algin, a gel used in baked goods and soft-serve ice cream. Brown seaweeds are craved as culinary delicacies in the Orient, where they also serve as fertilizers. And in California, giant kelps have been planted along rope rafts in an effort to increase the growth of marine life feeding on them.

Our cutaway of the Atlantic sea (Figure 11.16) displays three distinct kinds of seaweeds—all protoctists that are easily visible to the naked eye. The brown kelp *Laminaria* can be seen floating and attached at the lower left. The fleshy green algae *Codium* grows in branches. It can be seen covering the rocks and in the right lower foreground. The smaller red alga *Gracilaria* inhabits deeper waters than its green neighbors. It is depicted anchored to a rock on the lower left and supporting characteristic bulbs in the right center.

FIGURE 11.17
Various species of red algae from
a work by Ernst Haeckel (see
legend to Fig. 1.2).

Algal growth is most dense along rocky coasts where the organisms, exposed to sunlight, can get a foothold. The open ocean is usually devoid of seaweed except for some of the giant brown types.

▪ RED SEAWEEDS ▪

Most red seaweeds (also known as red algae or rhodophytes) are marine organisms that live attached to rocks. About 4000 species of red algae are known. Bulbous *Gracilaria* is shown in the lower left of Figure 11.16 and in Figure 11.17. Some species are a meter long and grow at depths of 180 meters in the ocean. It is the pigment phycoerythrin that allows red seaweeds to grow at such depths by capturing the longer-

152

wavelength light that penetrates that far. Plants and photosynthetic protists that possess only chlorophyll cannot tolerate depths below 100 meters.

The red seaweeds also contain other pigments, including the bluish phycocyanin, a key determinant of color in some of the blue-green bacteria. Indeed, red algae probably appeared when their nonphotosynthetic ancestors ate, but failed to digest, blue-green bacteria. Translucent, funguslike red algal ancestors probably became infected with cyanobacteria. In time, the cyanobacteria reproduced inside the host cells, becoming plastids. Molecular sequences of RNA in the photosynthetic organelles of red algae confirm this: they are so similar to those of certain cyanobacteria that the statistical likelihood of the two being directly related is overwhelming. The arrangement of the light-sensitive pigments and their membranes also is virtually identical in cyanobacteria and red algal plastids.

Red algae are second only to the brown algae in size. Leafy New England species are dried and eaten whole as "dulce," which can be bought in packages. Buy a bagful, dampen it with sea water and observe its cells with your microscope.

Other red algae are used in the manufacture of food and agar. Red algae are rapidly gaining in commercial importance. The necessary gels for distinguishing fragments of DNA with different base sequences in genetic engineering are made of agarose, a product most easily obtainable from certain red seaweeds that live in the Pacific Ocean. In addition, Japanese cuisine, both popular and healthy, often includes a diversity of rhodophytes.

▪ GREEN ALGAE ▪

Among the thousands of kinds of green algae are the ancestors to modern plants. Like plants, they are grass green in color, contain chlorophylls *a* and *b*, produce oxygen, and respire in both the dark and the light. Green algae differ from plants in their lack of embryo formation; nonetheless, many kinds of these "water plants" are known in both unicellular and multicellular forms. The green algae fall into two major groups: the diverse phylum of chlorophytes, which form swimming reproductive cells bearing undulipodia, and the gamophytes, the conjugating sort of green algae that never form cells with undulating cell tails.

You can easily collect green algae of many kinds. Look for scummy blooms on lakes and ponds, especially along the shore. When found, you can keep mixtures of green algae alive indefinitely in well-lit aquaria or large glass bowls. As the algae grow, periodically remove the excess green scum and keep the water level up with commercial spring water or tap water that has aged at least a day so that any chlorine gas it contains can escape. Growing green algae in this manner will attract associated microbes and provide you with plenty of material for microscopic observation, enough to last months.

Most green algae have a single large nucleus in each cell. They all can reproduce asexually, although many have quite complex sex lives. One type of gamophyte, *Spirogyra*, is filamentous. When *Spirogyra* feels sexy its filaments pair up side by side. Projections emerge and grow into conjugation tubes (Figure 11.18). Male cells slip through the connective tubes to the female cells, where together they form a dark mass in the female filament.

153

These fused cells form a thick, resistant cyst allowing them to survive the harsh season. Eventually the fused cells divide, reducing the number of chromosomes per cell in the process known as meiosis. After emerging from the hard, dark cyst, *Spirogyra* cells germinate, growing by mitotic division to produce fresh filaments.

Another kind of gamophyte are the desmids, some of which form colonies. Desmid cells, enveloped in mucilaginous sheaths, often are adorned with intriguing patterns. Their cell walls contain cellulose, pectinlike substances, and sometimes are impregnated with iron or silica. Their slow, creeping movement seems to be due to the secretion of mucilage through pores in the cell wall. (It is likely that actinlike protein filaments are also involved.) Desmids are composed of two half cells that mirror each other, the result of reproduction. They, too, sometimes reproduce sexually: the two half cells emerge from their shells and fuse outside. In some species, only the male leaves its shell and enters that of the female, where fusion takes place.

The other type of green algae, chlorophytes, are not directly related to each other, but are grouped together because of certain, easily observed similarities. Their cell walls, like plants, contain cellulose and pectin, or mannose or xylose derivatives linked to protein. Like land plants, green algae also tend to be sexual organisms. Unlike flowering plants, however, green algae do produce swimming sperm. A trend can be traced in green algae from isogamy, the production of equal-sized sex cells, to anisogamy, sex cells differentiated in size, as in species that form small moving sperm and larger, usually sedentary eggs. In the well-studied *Chlamydomonas*, the sex cells are isogamous but, although we have difficulty telling the sexes apart by looking, they don't make mistakes. A given cell (mating type +) only mates with one of the opposite mating type (mating type −).

Green-colored seaweeds, common on the beach and easily visible to the naked eye, also are a form of green algae. They appear to be multicellular organisms; some, like *Caulerpa*, are a meter long. In reality, *Caulerpa* is not multicellular. Because no cell membranes or walls form to define the cell boundaries, these algae are strangely crammed single cells. They are multinucleate masses with millions of nuclei, chloroplasts, and other organelles all enclosed by a single cell membrane.

Ray beings look like burrs, spurs, and stars. As many as several hundred long, straight spikes shoot out in all directions from their bodies (Figure 11.19). The rays, technically known as axopods, are the identifying characteristic among all members of this group.

The axopods have a variety of functions. *Sticholonche zanclea* uses them as oars to row the salt waters of the Mediterranean. *Acanthocystis* uses them to grab swimming prey—small protists or even animals. Stuck to the rays, the prey organisms die and are digested, their contents sucked down the rays into *Acanthocystis*'s cytoplasm.

Sticholonche and *Acanthocystis* have radial spines or spicules composed of strontium sulfate. Other ray beings have spicules made of silica. These are the strikingly beautiful radiolaria, masters of silica biomineralization. Radiolaria are not related to each other evolutionarily, but are composed of two distinct groups: the polycystines and the phaeodarians. Polycystine skeletons are opaline—made of hydrated amorphous silica, as are semiprecious opals. Polycystines also secrete siliceous hard skeletons. The biominerals begin as tiny, watery silica deposits associated with the internal membranes of the cells. The structure of phaeodarians, on the other hand, remains more enigmatic: they contain silica but also other, still uncatalogued materials.

FIGURE 11.19

Sticholonche, an axopod ray being that swims in the Mediterranean sea. (Drawing courtesy of J. and M. Cochon.)

Axopods increase the surface area of ray beings, allowing some to float and scavenge for food on the surface of the ocean. The rays also provide a surface for the accumulation of nitrogen and phosphorus in the nutrient-poor waters of the open ocean. In some ray beings, the rays are used for locomotion as the protists, looking like tumbleweeds, roll as they move. These locomotory axopod spikes have been studied under the electron microscope which has revealed their internal structure to consist of fine cylinders known as microtubules, often in elaborate arrangements. For example, a ray of *Sticholonche zanclea*, looked at in cross section through a scanning electron microscope, appears similar to a Chinese checkerboard, a giant hexagon of hundreds of microtubules linked together in triangular formations (Figure

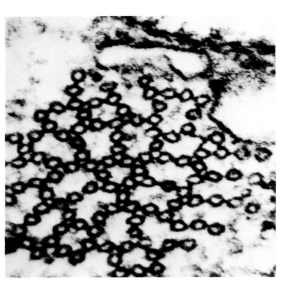

FIGURE 11.20

Ray being microtubules. Transmission electron micrograph shows microtubules from "oars" of *Sticholonche* linked in a hexagonal array. (Photo courtesy of J. and M. Cachon.)

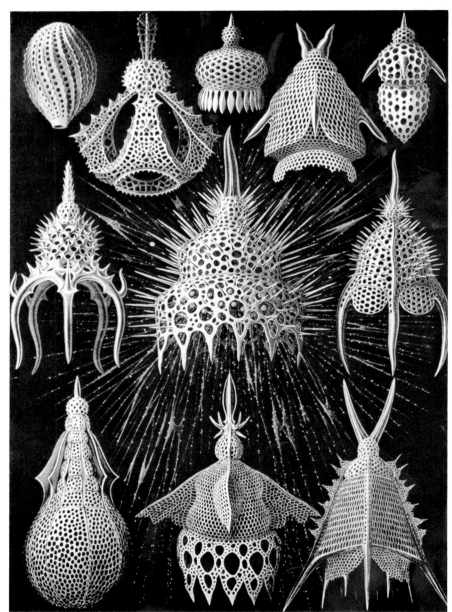

FIGURE 11.21

Spiny silicate shells of radiolarians, from a work by Ernst Hoeckel (see legend to Fig.1.2a).

11.20). Some ray beings have spherical "cages" arranged in repeating hexagonal patterns through which the rays protrude. Electron microscopic studies show that ray beings differ greatly in their fine structure. This seems to indicate that the tendency for making long rays or spicules remains part of the genetic structure of many protists, that rays have evolved independently in a variety of beings in marine and freshwater environments.

Another unusual characteristic of polycystines is their chromosomal makeup. While many protoctists have only one set of chromosomes per cell, and plants and animals generally have two, the polycystines contain several. Although no ray beings are photosynthetic by themselves, many polycystines nurture symbiotic yellow or green algae in huge numbers in the netlike material that really is an extension of their own body.

These spiny beings can be collected from offshore marine samples. (They fascinated the German biologist Ernst Haeckel (Figure 11.21).) But, while easily observed and collected, ray beings are hard to keep in captivity. The freshwater ray beings, or heliozoans, are easy to collect from the bottoms of streams and ponds where they are often stuck to rocks.

▪ FORAMS ▪

Forams (Foraminifera) are marine unicells; some grow to several centimeters in diameter (Figure 11.22). Free-swimming forams are ecologically crucial to the ocean food supply; they form the main food staple of many invertebrate animals. And forams themselves are voracious eaters of anything from ray beings and algae to ciliates, the larvae of crustaceans, and even nematodes. Forams usually live in the sand, where they attach to stones and other organisms. Many of them are symbiotic, housing diatoms, golden algae, or sea whirlers inside their bodies.

Many foraminiferans have sex, complex life cycles, and more than a single nucleus per cell. Most have elaborate skeletons pierced with pores or emit spikes at some time during their life cycle. The shells of extinct varieties of forams have intrigued and beguiled fossil hunters because of their fantastic shapes. Most famous are the nummulites or "coin stones." Beautiful 10-centimeter-wide nummulites were widely distributed 20 million years ago,

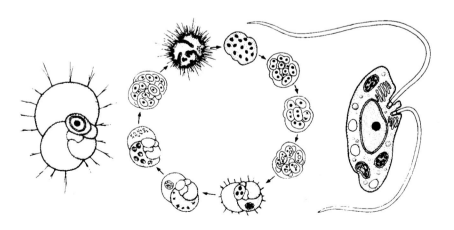

FIGURE 11.22

Life cycle of a foram: mating only occurs between nuclei in the same cell, reproduction is entirely separate from sex; an adult is shown at the left and a motile "swarmer" cell, highly magnified at the right. (Drawing by Laszlo Meszoly.)

157

forming rich layers of sediment on the ocean floor. At one time, there was even a suggestion that all beings on Earth were composed of nummulites, in part because it once was believed that primeval life came from the sea bottom. Despite the dubiousness of such propositions, it is true that 18-million-year-old forams are found throughout the limestone used to build the pyramids of Egypt.

Another quality of fossil forams is their use as markers in measuring the age of other fossils. Because they are so prevalent and distinct, geologists can match sediments from different parts of the world. Geologists seeking petroleum, for example, look for certain foram fossils as indicators of sediments covering oil deposits.

▪ CILIATES ▪

Ciliates are virtually omnipresent microscopic protoctists (Figure 11.23). They inhabit every kind of water imaginable, including ponds, streams, lakes, tide pools, oceans, and sulfur springs. Ciliates have two kinds of nuclei: macronuclei and micronuclei. Although not evident in the adults in all cases, ciliate cells tend to be covered with cilia during at least a portion of their life cycle. Even though most ciliates are small or large single-cells, they are composed of highly complex cell structures. *Sorogena*, such as the one shown here, are multicellular and look like mold to the naked eye (Figure 11.24).

Over 8000 species of ciliates have been described in the scientific literature. Nearly all scientists who study ciliates feel that many more species remain to be discovered, especially in the tropics. Ciliates have been found on the gills of every sea urchin carefully studied by marine zoologists.

Just as the detailed morphology of a flower reveals its origins, the finest structural detail of the ciliate surface is a key to its taxonomy and evolutionary history. The electron microscope has revealed three basic groups of ciliates: postciliodesmata, which have cell tail arrangements with "postciliary fibers;" rhapdophorans, which lack such fibers; and cyrtophorans, which reconstruct their entire cell surfaces prior to cell division every generation.

Ciliates have features that are generally attributed to animals. Some are predatory by way of extensible cords with poisoned darts at their ends, looking remarkably like baited fishing rods. Missiles are shot and impale unsuspecting prey. After the prey is caught, it is reeled in.

Although all ciliates are capable of reproduction without sex, many ciliates have sex lives. The terms male and female make little sense in most ciliate couplings because the pairing organisms look identical. Furthermore, ciliate sex is very different from animal sex. Ciliates don't make baby ciliates. Sex, to ciliates, is the production of extra nuclei and the subsequent exchange of these nuclei, each one containing a set of thousands of genes. In the case of the familiar ciliate *Paramecium*, if it is mature and ready for sex and a potential partner is not available or is taken away the extra nuclei will fuse with themselves in a kind of self-sexuality, called autogamy.

Nearly all ciliates have mitochondria and very few are capable of photosynthesis. One is *Paramecium bursaria*, which you can easily collect near the

FIGURE 11.23
Glaucoma chattoni ciliate seen with scanning electron microscope. (Photo courtesy of Prof. Eugene Small.)

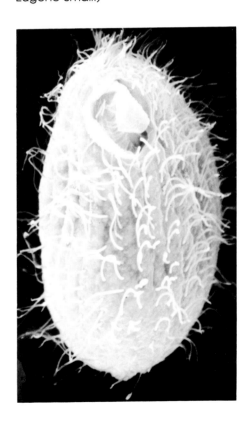

surface just off the shores of ponds. It is always bright green and thrives in nutrient-poor, sunlit waters. Inside the paramecium, the green alga *Chorella*—a symbiont protist—"fixes" carbon dioxide taken from the air into amino acids and other food stuffs. *Paramecium bursaria* can be induced to shed its green inhabitants, which it neatly contains in membrane-bound sacs, but this is difficult. If placed in darkness and treated with certain harsh chemicals that inhibit oxygen-producing photosynthesis, *Paramecium bursaria* will expel or digest its green symbionts. It will neither enter a sexual encounter, nor can it—unless force-fed—live in the light without its green algal partners.

The unit of structure of all ciliates, the part from which information about evolutionary relationships can be deduced, is the kinetid (Figure 11.25). (Although the existence of kinetids can be seen with a light microscope, electron microscopic studies are required to distinguish between different kinds.) The kinetid is a unit that at the minimum contains a kinetosome and an undulipodium. Sometimes a kinetid has two kinetosomes and two corresponding undulipodia, or two kinetosomes only one of which has an undulipodium (the other is called the "barren kinetosome"). In no case in nature is an undulipodium ever found without a kinetosome: this is because undulipodia always develop from kinetosomes. Kinetosomes, found at the base of undulipodia, have a curious mode of generation. In certain experiments they have been shown to reproduce directly from other kinetosomes by a complex process despite the absence of a nucleus elsewhere in the cell. In other cases they suddenly appear from what looks like fuzz. The idea that self-reproducing kinetids are the legacy of once free-swimming spirochetes is now being tested.

Ciliates are found in all water habitats all over the world. However, they often are overlooked because many form cysts that are resistant to starvation and water loss. To observe their transformation, try making your own "hay infusion." Add about a half cup of water and two or three dry peas from a soup package to a handful of dried grass or soil, and ciliate cysts should germinate and the ciliate population begin to grow.

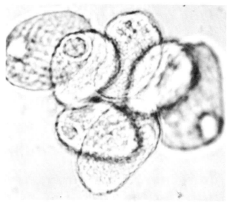

FIGURE 11.24
Sorogena, a ciliate that forms multicellular stalks composed of hundreds of ciliates. The round bodies at the top are cysts. Once in soil they will germinate to form swimming ciliates. (Photo courtesy of Professor Lindsay Olive.)

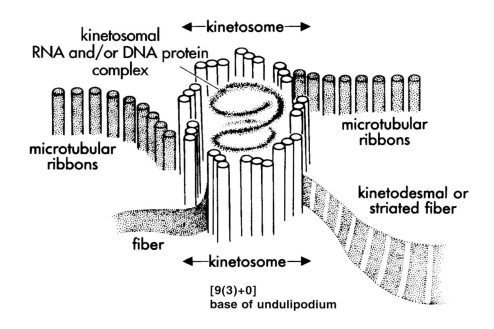

kinetosomal
RNA and/or DNA protein complex

←kinetosome→

microtubular ribbons

microtubular ribbons

kinetodesmal or striated fiber

fiber

←kinetosome→

[9(3)+0]
base of undulipodium

FIGURE 11.25
Structure of a ciliate kinetid (composed of microtubules). (Drawing by Laszlo Meszoly.)

159

To humans, the spore parasites (apicomplexans) are the nastiest protists. They are animal parasites, especially notorious as infectors of human blood. The best way to safely acquaint yourself with these microbes is by ordering a fixed and stained preparation from a biological supply house (see our list). Very few people have seen spore parasites divide while alive because they are so small and because specialized training is required to handle them.

In some ways, these tiny protists have life cycles more complex and intricate than humans. Like many algae and plants, they alternate generations: they spend much of their lives in haploid states with a single set of chromosomes, and a smaller part of their lives in the diploid state with two sets of chromosomes present in the nucleus of each cell. Apicomplexans reproduce sexually, but they also reproduce by a special process known as schizogeny: a series of rapid, mitotic cell divisions, in either the haploid or diploid state, that produces infective spores. During sexual reproduction, a whipping, spermlike cell fertilizes a larger female cell to produce the oocyst, which is relatively large and resistant to heat, dryness, and radiation. The infectious package usually is eaten unintentionally by the host and, once inside, schizogeny takes over to invade the cells of the victim.

Apicomplexans get their name from the apical complex, a special arrangement of microtubules and specialized organelles used as injection devices at the apical end of the cell. Because this structure is so distinctive, it is very likely that all apicomplexans are directly related.

Apicomplexans cause a wide variety of diseases. *Isospora hominis* causes dysentery in man and parasitizes livestock and fowl. *Eimeria* also causes animal infestation. *Plasmodium*, the most infamous apicomplexan, is the culprit responsible for malaria (Figure 11.26). Unlike the aforementioned ailments, which occur after oocysts are eaten, malaria occurs when infected *Anopheles* mosquitos bite people. The microbe lives in the gut of the insect and then commutes to live in the blood of humans when they are bitten by the insect. This required alternation of habitats, from mosquito to human, is very specialized. Inside human blood, *Plasmodium* feeds on hemoglobin, which gives the apicomplexans the protein their diet demands. Schizogeny of *Plasmodium* parasites occurs in waves, accounting for the periodic outbreaks of fever in malaria victims. But *Plasmodium* is only one apicomplexan; many more of these highly complex protoctists with their own alien lifestyles exist. Some already discovered have not yet been studied very well. Others, more the province of medical researchers and veterinarians than biologists, are still so complex and secretive that they remain virtually unknown.

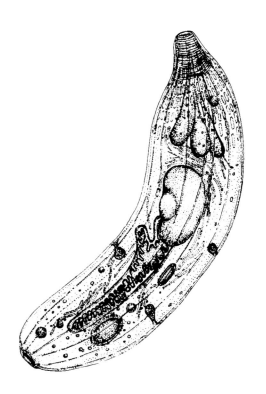

FIGURE 11.26

Plasmodium, an apicomplexan that causes malaria. The cell's "apical complex" is a highly structured modification of the anterior portion of the cell that insures sticking to and penetration of our own red blood cells. (Drawing by J. Steven Alexander.)

■ POLAR THREAD PARASITES ■

In the early literature of protozoology the polar thread parasites were classified with the apicomplexan spore parasites as sporozoans. Technically known as cnidosporidians, the polar thread parasites are really a motley group of microbes united only by their mutual production of some kind of filament or thread at the ends of their cells. Although they do form

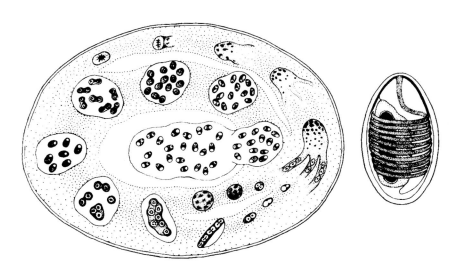

FIGURE 11.27
Life cycle of a microsporidian polar thread parasite as it would occur in an animal such as a flounder. After infection by a polar thread parasite, the host nucleus undergoes cell division. Many parasitic plasmodia grow and then fuse to develop into numerous mature polar thread parasites, which can then go on to infect other cells. (Drawing by Laszlo

spores, the polar thread parasites lack the apical complexes and infectious spores of the spore parasites. Instead, their spores represent resting stages between bouts of parasitic infection.

There are three main kinds of polar thread-forming microbes: the injecting microsporidians, the anchoring myxosporidians, and the hooking actinomyxidians. The microsporidians live inside the cells of the animals they attack, where they reproduce, often with great fecundity (Figure 11.27). Like living microbial needles, microsporidians inject their infective reproductive part, called the sporoplasm, through a narrow tube that develops from the polar thread. Polar thread parasites are not nearly so vicious as spore parasites, however, and many microsporidians live in the tissues of their hosts without causing the least damage. Others, however, cause harm to many animals, from starry flounders to silk worm larvae.

Polar threads are used in an entirely different fashion in myxosporidians. Rather than inject infectious agents, myxosporidians anchor themselves to host tissue. Their polar threads are encapsulated in a cnidocyst. Because the cnidocysts of myxosporidians bear a striking resemblance to the nematocysts of coelenterate animals, it is quite possible that in a reversal of the usual evolutionary direction these protoctists evolved from animals. But since they do not form hollow, pre-embryonic blastulas as coelenterates do, myxosporidians must be considered protoctists.

Myxosporidians are associated with a variety of diseases, especially in fish. In salmon, they cause twist disease. They also infect the European barbel, a freshwater fish. Myxosporidians have spores with many nuclei, but when they penetrate the skin and make their way to the intestines of their hosts they produce uninucleate cells called amebulinas. From there, the myxosporidians expand into tissues and organs, especially hollow ones such as gills, bladders, and bones.

Finally, there are the actinomyxids, poorly known organisms that form parasitic associations with invertebrate animals, particularly segmented simple worms such as oligochetes and annelids. The actinomyxids attach to their hosts by way of their thread-containing spores. These are partitioned into two hooklike sections, also called cnidocysts, but which probably are not related to the cnidocysts of myxosporidians.

The best way to see polar thread parasites is to skim the whitish fuzz from the sick-looking fish of a poorly tended aquarium. Aquaria in general provide wonderful sources for entries into the microbial world. "Mold" from "moldy fish" generally isn't mold at all but protoctist polar thread parasites. To examine them just take a tiny tweezer-full and look at it under your microscope. Don't worry, though; the polar thread parasites that infect fish can't grow on or hurt you.

▪ SLIME NETS ▪

The slime nets, composed mainly of marine microbes, are colonial organisms that move within their own secreted slime (Figure 11.28). Individual cells glide up and down in a network of slimy, elastic tubes—the so-called net plasmodium, which can itself slowly creep about. Slime net microbes grow on and infect such marine grasses as eel grass, *Zostera marina*, and various kinds of seaside algae. Because eel grass is a rich component of Atlantic coast ecosystems that nurture clam and oyster beds, slime net infections of *Zostera* can lead to shellfish destruction. The best-studied slime nets are the labyrinthulids. You can collect *Labyrinthula* if you learn to recognize it as clear mucous lines on eel grass or on *Spartina*, a type of salt marsh grass. Microscopic examination reveals that the slimy substance

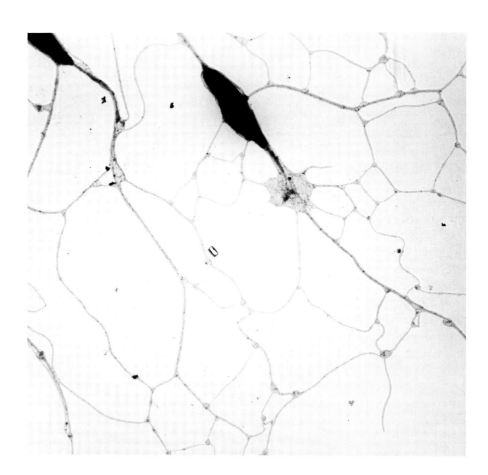

FIGURE 11.28

Labyrinthula, a slime net. The threadlike structures in this light micrograph are the slime trails within which the small coccoid cells move and grow by division. (Photo courtesy of Frank O. Perkins and the Canadian Journal of Botany.)

is really a highly developed transportation system for slender rodshaped cells with nuclei. The cells move back and forth, apparently at random. They move quite fast for their size and are transported not by movements of their own so much as by a mysterious mechanism of the tracks. This is deduced from the individual slime net microbe's inability to move outside the slime tracks.

Labyrinthulids eat by sensing food, moving towards it, and then secreting digestive enzymes. Food molecules resulting from enzymatic breakdown provide nourishment for the slime net colonies, which absorb them from the outside through the net system. Reproduction and growth ensue as each cell divides by mitosis, inside the net. New cells presumably secrete slime as well as enzymes to expand the slime network. Sexual processes probably occur in slime nets. Some labyrinthulid cells have been observed with the heterokont design: that is, cells with two undulipodia, one directed forward (which has small sidewise projections called flimmer) and the other trailing.

Beside *Labyrinthula*, there are three other slime net genera. Furthermore, about 30 species of obscure marine and estuarine organisms called thraustochytrids, long thought to be peculiar forms of fungi, are now considered to be relatives of slime nets. You may be able to get thraustochytrids to reveal themselves by taking a jar of sea water and adding to it yellow, pine tree pollen taken from a pollen cone. Check the clear or whitish "mold" that appears after a few days. Thraustochytrids are like slime nets in that they also produce an extracellular slime track devoid of nuclei, mitochondria, and other organelles. In both types of slime nets, a special structure called a sagenogenosome produces the extracellular network. Thraustochytrids can be distinguished from labyrinthulids because they don't actually move inside the slime network.

▪ CELLULAR SLIME MOLDS ▪

If science fiction writers competed to dream up the most unusual organisms to plausibly inhabit alien worlds, an entrant might win by submitting *Dictyostelium*, a cellular slime mold already living on the Earth. This slime mold begins its bizarre life cycle as a mass of spores suspended on a slender stalk (Figure 11.29). Under certain conditions, the spore case breaks and amebas emerge. Although the amebas look very much like other nearby amebas, they will aggregate if deprived of nutrients. When placed on a small dish of growth medium that is streaked with food bacteria, *Dictyostelium* amebas gorge themselves until there are virtually no food bacteria left. Satiated and still dividing, the amebas reproduce until they cover most of the plate. At this point they begin to call out to each other in chemical code. They converge, attracted to each other by cyclic adenosine monophosphate, a type of energy molecule used in cellular metabolism and motility. The excited, aggregating amebas climb on top of each other, forming a "slug"—a mound of protoplasm with a nipple on top. In at least one species, this structure then collapses sideways and deliberately slides toward the light before continuing its growth. The slug attenuates, stretching until it is a slender stalk supporting a spore case. Under appropriate condi-

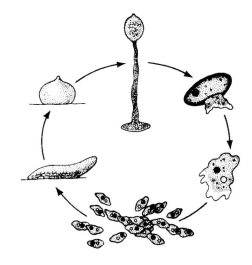

FIGURE 11.29
Life cycle of a cellular slime mold. Individual amebas coalesce to form first a migrating slug and then a cyst-bearing stalk. (Drawing by Laszlo Meszoly.)

tions, the spore case will break, releasing individual spores each of which contains a new ameba. The multicellular organism is in a sense altruistic: in the interior of the aggregated mound a core of amebas are surrounded by a sheath that fills with a fluid, killing the amebas—all in order to support other amebas, which climb over the sacrificed cells to make the stalk grow. The survivors of the narrowing stalk climb over others to the top and form the spores representing future dictyostelid generations.

Dictyostelids, and other even more obscure cellular slime molds called acraseans, can be found in temperate woodlands on rotting logs and in damp soils over which they slowly migrate, leaving behind a track of slime derived from their collective sheath. They are also found in fresh water. A few acraseans grow on animal dung, but perhaps that habitat chemically confuses them because there their stalks and spore heads combine and are not fully formed. While most cellular slime molds inhabit forest soils, some have moved into gardens and lawns; one newspaper report described the amazement of golfers encountering the unusual life form spreading visibly across their golf course.

Slime molds combine traits of some of the more familiar life forms: they move, eat, and metamorphose like animals. They form spores that resemble ameba cysts and grow upward like plants. They inhabit decaying plant material like fungi. But the 19th century German biologist Ernst Haeckel intuited that such biological ambiguities really belonged to none of these three kingdoms but to an ancestral one of their own, which he created for them and called the kingdom Protista.

These slime molds never lose their cellular organization: the amebas in the aggregating slime, the migrating slug, and the top of the stalk consist of complete cells separated from each other by encircling cell membranes. Reports of sexuality and cell fusion in these species have been questioned; they may be isolated incidents of cannibalism rather than mating. In any case, regularized sexual patterns have not been well documented. Became they eat potentially contaminating bacteria, these organisms tend to be easy to collect and to keep. (For detailed instructions, look up *Dictyostelium* or *Acrasia* in the ATCC Catalogue or in Olive's book (1975).) Watch for moldlike slime that upon closer examination appears to be collections of tiny grey or yellowish stalks. These will tend to be either cellular or noncellular slime molds. See if you can figure out which.

▪ NONCELLULAR SLIME MOLDS ▪

The noncellular slime molds are similar to the cellular slime molds in many respects with the notable exception that they can form motile, slimy masses, called plasmodia, that are not divided into individual cells. At certain stages, they form cells with undulipodia and engage in sex but they are distinguished by the unmistakable plasmodial stage during which the moving slime mass seeks out food (Figure 11.30).

Also known as true slime molds, the noncellular slime molds transform into distinct adult forms that correlate to feeding and spore-releasing phases. If food is near, individual cells switch easily from swimming cells with one long and one usually very short whiplashing tail into tailless amebas that grow by nuclear rather than cellular division into amorphous, multi-nucleated plasmodial masses. This mass can produce swimmers again at any time. Both swimmers and amebas can mate, although not with each other. However, they can also reproduce asexually. In fact, you can induce reproduction in these organisms simply by breaking the decaying logs on which they are found into pieces.

Noncellular slime molds feed by creeping over fallen logs, bark, and other decaying vegetable matter. As it creeps forward, the liquid being looks like a rounded, constantly changing, rubbery leaf with veins fanning out towards the back. Cellular fluid courses through these veins, toward the rounded front, and back again in a pulsating manner. A trail of undigested bacteria and fungi reveals the path it has travelled.

Some of these slimy beings can be shaken or cut into pieces and each piece will rejuvenate into new throbbing slime. Analyses of the most easily collected of these slime molds, the bright yellow *Physarum polycephalum*, have yielded myxomyosin, a contracting protein similar to the actinomyosin found in our muscles. Indeed, were it not for its habit of restlessly dismantling and reassembling its longitudinal fibers rather than keeping them in a set arrangement as do the cells of animal muscles, a *P. polycephalum* plasmodium could be thought of as a huge, free-living, everchanging muscle cell.

Examples of plasmodial movement are legion. In one laboratory, noncellular slime mold spores placed in an envelope within a drawer escaped within hours, germinating in the damp air. The plasmodium then crawled across the room to form a film on a wooden bench, later surprising a student who touched what he first took to be water. On the outside of the bench drawer, where weather conditions for the plasmodium were more dry, bits of slime mold flaked off into black dried-out bits called sclerotia. The student scraped a tiny amount of the black flaky bits into another envelope and sent it to friends. Slime mold fans share their organisms in this fashion.

Unlike such surface-creeping slime molds, the common *Stemonitis* lives buried within decaying wood. Its presence is revealed by creamy droplets that are seemingly sweated out upon the bark. These collect into an inch-high

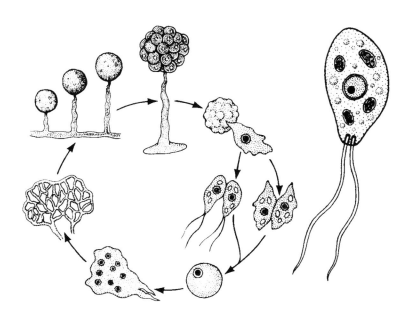

FIGURE 11.30

Life cycle of an acellular (noncellular) slime mold. A mastigote loses both its undulipodia and grows into a huge plasmodium, which later grows a cyst-bearing stalk. (Drawing by Laszlo Meszoly.)

165

FIGURE 11.31

Ceratiomyxa fruticulosa, a non-cellular slime mold growing on a log. (Photo by William Ormerod.)

cushion and then further condense into columns. The columns become fibrous and purple-colored, and later nearly black as they dry out and their branching spore heads, filled with dark spores, separate from each other. The spore heads are now so delicate that the slightest breeze or touch disperses the spores, leaving only the fragile skeletal stalks topped by brittle skeletal filaments, called capillitia, that once held the spores, and an underlying sheet of dried silvery slime. This amazing transformation takes fewer than 24 hours. The skeletal "blossoms" vary from species to species; some even have cogs and spirals sculpted into their walls. The capillitia puff up like cotton candy, offering their spores to gentle winds for dissemination.

Despite their literally creepy qualities, some noncellular slime molds are truly beautiful. Rare colonies of *Ceratiomyxa fruticulosa* cover decaying vegetation with sprawling miniature forests of brilliant yellow branches (Figure 11.31). The young *C. fruticulosa* becomes white upon maturity as it prepares to make its propagules. Other slime beings may be orange, brick-colored, or purplish-white. And, while no slime mold is ever photosynthetic, some thrive and construct structures on logs as intricate as coral reefs—structures that can only be described as microbial gardens.

▪ PARASITIC SLIME MOLDS ▪

The parasitic slime molds are mysterious, tenacious, subterranean organisms. None of their life cycles has ever been observed from start to finish, despite the fact that they have been studied with labor and diligence (Figure 11.32). They are called plasmodiophorans or plasmodiophorids by scientists who are trying to figure them out.

Parasitic slime molds have been known for thousands of years as club root, causing monstrous overgrowths in the roots of cabbage. They also cause deformations in other members of the mustard family, such as brussels sprouts, cauliflower, and kale, as well as in turnips and radishes.

Although mysterious, these slime molds are meticulously adapted to their role as parasites. Unlike the other, free-living slime molds we have encountered, the parasitic types don't throb and creep over surfaces, eating bacteria along the way. These slime molds simply snuggle, motionless, in plant tissue, from which they absorb the juices they use as nutriment to grow more nuclei, more cytoplasm, and more cell material.

Parasitic slime molds are thought to grow into plasmodia from swimming cells with a short undulipodium in front and a long trailing one pushing it along. Reminiscent of the soil amebomastigotes, from which these parasites may have evolved, these swimmers are released from infected plant tissue into the surrounding soil where they may remain swimmers or transform into amebas. The swimmers with their whipping tails invade the cells of plant roots where they undergo nuclear division in the manner of other slime molds and grow into subvisible plasmodia.

Inside the cabbage or cauliflower, the intracellular plasmodium forms a coating and then cleaves off into fragments. Nuclei in the fragments duplicate and divide at least twice to form at least four new swimmers. The restless, hungry swimmers are discharged into the soil through a special exit tube in the plasmodium. This second batch of swimmers, it is surmised, may have sex, although no one has yet documented it. It has been postulated that the fused cells again enter plant roots.

Parasitic slime molds have a long tube and a bulletlike appendage that allows them to vigorously attack their vegetable hosts. The plants are doomed as the parasite sucks their bodily juices from inside. As the parasites grow in the cabbage tissue, they swell the roots, forming hideous swollen tumors on them, while at the same time stunting the growth of leaves. One can observe parasitic slime molds on the foods we share with them, including potatoes, wheat, barley, oats, rice, and peanuts. The easiest way is to look for and at swellings on cabbage or other plant roots. One easily susceptible host is *Brassica peteinensis,* Chinese cabbage.

The slime molds' zoospores can last for years in soil so that if half a hemp or sesame seed is added to a small amount of soil along with distilled water,

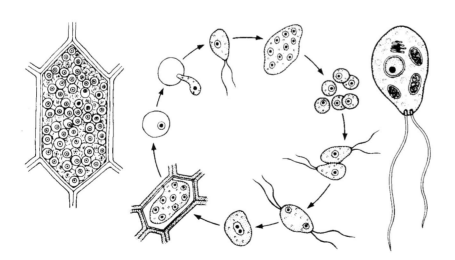

FIGURE 11.32

Life cycle of plasmodiophoran parasitic slime mold. This kind of organism develops its plasmodium inside cabbage and other plant cells. Cabbage cell filled with plasmodiophoran is shown at left, single motile cell at right. (Drawing by Laszlo Meszoly.)

167

growth of plasmodiophorids is likely to ensue. If you suspect you have them, add the soil–water mixture to small pots in which some sort of mustard plant, such as cabbage, kale, brussels sprouts, or kolrabi, has been planted or is growing. If the plant develops abnormal growth patterns and monstrous swellings, you probably have parasitic slime molds, which you can then see by observing the swollen plant tissue under a microscope.

▪ CHYTRIDS ▪

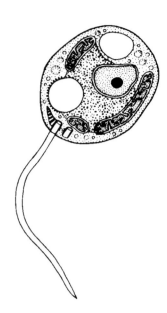

FIGURE 11.33

Chytrid swimmer cell. The whole chytrid is much bigger and looks like fuzz to the naked eye. (Drawing by Laszlo Meszoly.)

R elatively large multicellular protoctists that traditionally have been considered fungi, chytrids may become visible as infections on potatoes. Their name derives from the Greek *chytra*, meaning an earthen cooking pot. Indeed, in certain stages of their development, many of them do resemble cooking pots with their dark and vessellike forms rounded at one end and flat across the top.

Superficially, chytrids are very similar to fungi except that, unlike any true fungi, they produce swimming cells with whipping tails (Figure 11.33). Indeed, their habit of gaining nutrition by extending roots and threads, secreting enzymes, and absorbing nutrients through enzymatic breakdown is so like fungi that to find out about them from books you need to look them up under fungi or, more recently, as "protistan" or "protoctistan" fungi. Another compelling connection with fungi is their method for making the amino acid lysine. The chytrids may in fact be the protoctistan survivors of the ancestral lineage that evolved into fungi.

Chytrids, like fungi and also insects, have cell walls composed mostly of chitin. Although sex is not required for reproduction, chytrids do indulge in sexual activities. Females of the species *Allomyces* secrete a hormone called sirenin, named after the attractive "sirens" of Greek mythology. The female attracts male cells, which swim toward and fuse with the unicellular females. Male and female cells may join each other as they swim; at first glance, they appear to be identical in structure. While some chytrid species display a tendency towards altering the primeval unity between male and female cells, there is at least one swimming cell—assessed to be the male—in all chytrid species.

After fertilization and fusion chytrids become adults, many within cells of both plant and animal hosts. Some adult chytrids have unicellular bodies. More complex chytrids feed deep inside the tissues of their hosts while distributing their reproductive organs upon the outside surfaces of their unfortunate victims. Fused cells grow into tough, protective structures that produce either new feeding structures or zoospores that germinate into new feeding structures.

From the developmental point of view, some chytrids are remarkably versatile: their fertilized cells, unlike the fertilized eggs of animals, develop into one of a variety of adult types, depending on environmental factors. *Blastocladiella emersonii* can develop into an ordinary colorless body, a resistant dark feeding structure, or a tiny body that releases only a single swimming cell. These three options are exercised under differing conditions associated with local quantities of moisture, food, and carbon dioxide gas. Like a science fiction character that conveniently assumes human form, *Blastocladiella* changes its shape and life cycle to meet the contingencies of

168

existence. Some chytrids live inside pollen grains, others invade unicellular green algae that float on the surface of lakes, thereby depriving whole ecosystems of the photosynthetic food upon which they depend. Freshwater chytrids may live on twigs of ash, birch, elm, or oak that have fallen into slightly alkaline pools and ponds. The wart disease of potato is associated with the chytrid *Synchytrium endobioticum*. Widely spread in Europe, *S. endobioticum* causes dark brown, cauliflower-shaped growths that deform potatoes. Although in other parts of the world potatoes have built up a resistance, the disease continues to be a serious problem, especially in Europe and most recently in Newfoundland, Canada.

▪ TINSEL CHYTRIDS ▪

Like the parasitic slime molds, the tinsel chytrids (Figure 11.34) remain elusive. In part this is because, like all parasites, their strict habit of living off of other organisms inhibits scientists from observing all parts of their life cycle. Some tinsel chytrids, for example, spend their entire lifetime parasitizing pollen grains of evergreen trees. You can "fish" for them in soil using pollen grains as "bait." Scientists coax tinsel chytrids out of the soil by exposing soil water, which contains them as dormant cysts, to pollen grains.

The tinsel chytrids, technically known as hyphochytrids, used to be considered fungi. But they are not true fungi because they form an undulating tail at one stage in their life. Each tinsel chytrid bears a single undulipodium with flimmer. Although flimmer can be seen only under a transmission electron microscope, the presence of it in tinsel chytrids separates them from the plain chytrids that have no "tinsel" on their undulipodia. The detailed characteristics of the undulipodia—and the nature of their insertion into the cell—are thought to be far more dependable as taxonomic markers than, say, multicellular shapes. These characteristics, however, are only barely visible even at high power with a good light microscope.

Tinsel chytrids represent a small group of microbes, possibly unrelated to each other, that superficially resemble the much more well-known chytrids. Some live in the soil; many others are freshwater microbes that live in the tropics. They feed on fungi, algae, plants, and insect larvae. As in the cells of euglenoids and slime beings, the undulipodium is located at the front of the cell. These cell whips may turn like spirals or wave back and forth in order to propel the tinsel chytrid "zoospore" through the water. The swimming stage of tinsel chytrids alternates with an absorbing, funguslike stage. Transformation of cells from swimmers cells into a multicelled absorbing stage takes place directly and, as far as it is known, without any sort of sex. Like true fungi, tinsel chytrids in their absorbing stage will produce slender strands that release enzymes, which break down tissue or decaying organic matter. Live or dead, this matter is then sucked up and nutrients are absorbed through the weblike collection of strands.

The bodies of some parasitic tinsel chytrids form inside the tissues of their unwilling hosts. Reproductive structures consist of a series of rootlike tubes (rhizoids, or hyphae) used for feeding and anchoring onto surfaces. Each reproductive structure releases whipping swimmer cells through an exit

169

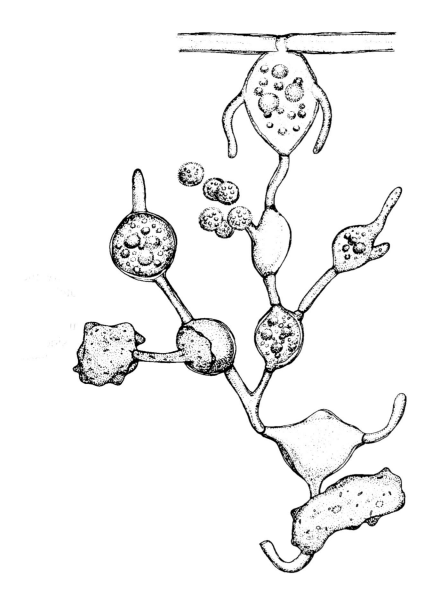

tunnel. In the species *Rhizidiomyces apophysatus*, the swimmer cells seek out
certain water molds (egg molds, or oomycotes), another kind of funguslike
protist that forms a whitish fuzz on plants, fish, or fish eggs. The swimmers
attach to the water mold cells and become round by pulling in their undulating, flimmered tails. Then they "germinate," sending out tubes into the water
molds. These tubes grow into webs or nets of tiny roots swelling up near the
host cell's surface. The swelling, baglike structure fills up; an exit tube
consisting of a raised bump with a hole in it forms at one end and is populated
by the divided nuclei of tinsel chytrid cells. Protoplasmic globules containing
many nuclei are ejected through the holes and transformed by a process of
cleavage and differentiation into a new swarm of rapidly swimming tinsel
chytrids. Despite their parasitic lifestyle, it is thought that tinsel chytrids are
beneficial because they do not infect us or our foods but rather attack some of
our enemies, such as the egg molds, which cause potato blight, the blue mold
of tobacco, downy mildew in grapes, and other crop-destroying diseases.

The egg molds (oomycotes) are the best known of the water molds (Figure 11.35). You can easily pick them up as fuzzy white material on decaying seeds, leaves, and other pond vegetation. They can be distinguished by their round "eggs," which, however, are not fertilized by their spermlike zoospores. No sperm are developed in them; rather, the egg mold is distributed by sexless zoospores. These zoospores retract their undulipodia and grow into the fuzz that makes up the body of the watery egg mold. The male fuzz grows over to the female fuzz, apparently attracted by the large round egg inside. Both male and female parts can grow from the same fuzzy mold. The male–female touching leads to the production of special "fertilization tubes" that serve as a conduit through which are passed the nuclei of male cells. If conditions are wet after fertilization, a structure grows that becomes filled with the swimming zoospores. Eventually the swimmers emerge through a hole and dart away to form new patches of egg mold fuzz. In some oomycotes, and especially if conditions are threatening, the following may occur: after moving through the tunnel, male nuclei come upon egg cells, known technically as "oospheres," which lie inside the walls of the female reproductive parts. The male nuclei fertilize these oospheres and the fused cells develop into dark spores resistant to microbial drought and famine. Patterns at the surface of these resistant spores, now called "oospores," may be used to distinguish among egg mold species.

Like the chytrids and tinsel chytrids, egg molds have cells with waving undulipodia. Their swimming cells have one small, forwardly directed undulipodium with flimmer, and another propulsive one that trails, whiplashing the rest of the cell through the water. Also like fungi and other funguslike microbes, oomycotes feed by extending a network of enzyme-secreting threads through which they digest their dinner.

The most infamous oomycote is certainly *Phytophthora infestans*, devastator of countries, crops, and the populations that depend on them. Mass emigrations of Irish and German people to North America can be traced to *P. infestans*, which caused the potato blight of 1845–1847. The blight totally ruined the single-crop economy of Ireland, causing millions to die of

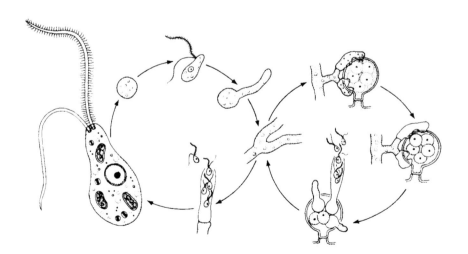

Figure 11.35
Life cycle of egg mold (oomycote). You need a microscope to see primary or secondary zoospores. Swimmer cells (left) turn into zoospores which germinate into structures that look like fungal hyphae. Fertilization leads to the development of new swimmer cells. This organism looks like fuzz to the naked eye. (Drawing by Laszlo Meszoly.)

starvation and millions of others to flee their country, many to the United States.

The oomycotes are distributed in lakes, rivers, and soils. They cause economically important afflictions such as the moldy fuzz that kills fish, white rusts (white blisters on the leaves and stems of cabbage, radishes, and other crucifers), downy mildews (potato blight, molds of tobacco, grapes, and other food plants), and "damping off" (a disease affecting young plants). These various maladies are spread in a variety of ways. Spores of white rust, caused by the species *Albugo candida*, are released when blowing wind or rain lands on fuzzed-over leaves. Zoospores with their twin cell whips do not swim but are cleverly housed in aerial structures. The same is true for the downy "mildews," whose spores disperse when wind blows past branched spore-containing structures. Perenosporales*, which spread rapidly through dense crops of a single plant species, are also wind dispersed. Water mold spores must germinate in the water, while spores from the oomycotes that cause "damping off" swim away when rain water saturates the soil.

People can also be important to oomycote dispersal simply by importing and exporting diseased plants. *Plasmopara viticola,* for example, which causes downy mildew of grapes and which is native to North America, was inadvertently brought to Europe in 1870s. French grape vines had no immunity to this pathogen and might have been irreparably destroyed were it not for the introduction in Bourdeaux of the first practical "fungicides." (These chemicals, lethal to the troublesome protoctist *Plasmophara,* should perhaps be renamed "protoctocides.")

You can trap egg molds with a cage of plastic-coated wire mesh containing an apple, an orange, crab apple, or rose hips. Find a place where there is shallow water, either stagnant or slowly moving over mud. Suspend the cage just below the water's surface for about 10 days. Remove the fruit at the end of this period and examine the pustules with a low-power microscope. You should be able to see some egg molds. Expect other inhabitants of the microcosm to be present at the feast, too. Examine the white fuzz over the course of a few days. You'll be able to distinguish the oomycotes by their oospores inside the "oosporangium."

---

* Perenosporales are a group of obligate symbionts of plants, some of which cause epidemics on crops. Each oogonium contains a single egg. Common names for these protoctists include "damping-off" (genus *Pythium*), "white rust" (*Albugo*), "late blight" of potato *Phytophthora*, "blue mold" of tobacco, and "downy mildew" of grape.

FIGURE 12.0

Common basidiofungi of a temperate forest: Can you identify *Lepiota procera, Heterobasidium annosum, Phallus impudicus, Scleroderma aurantium, Gomphidius glutinosus, Lactarius fuliginosus, Lycoperidon perlatum, Polystichus versicolor, Amanita muscaria?* (Drawing by Christie Lyons.)

FUNGI

▪ INTRODUCTION TO FUNGI ▪

Television commercials promoting cleansers, antiperspirants, and disinfectants assume microbes are universally dangerous, disgusting, disease-causing germs. All microbes should, according to the pseudoscientific demonstrations of the electronic soapbox, be neutralized, deodorized, and otherwise zapped into oblivion. Fungi ruin your bread, rot your house, and make your feet smell. Even the 18th century French botanist S. Veillard is reported to have exclaimed in a fit of ill temper that, "Fungi are a cursed tribe, an invention of the devil, devised by him to disturb the harmony of the rest of nature created by God."

Yet fungi (Figure 12.0), our last kingdom of microbes, have composed a symphony of evolutionary delights to which even the most diehard adman could not turn a deaf ear. They produce the alcohol in champagne, wine, and beer; from them come the mind-altering drugs psilocybin and LSD; they raise and texture bread, they ripen brie, camembert, thenay, troyes, and vendome cheeses; they flavor soy sauce and miso, and roquefort, gorgonzola, and other blue cheeses; they include not only all the common, supermarket-variety mushrooms, but also the prized truffles, the gourmet chanterelles and morels. Inside plant roots, fungi perform a service to all life by scavenging nutrients from the soil, adding to the nutrition and splendor of flowers and trees. Indeed, without the intimacy of fungi in the roots of trees there would be no wood.

Some fungi save people's lives by manufacturing penicillin and other antibiotic medicines inside their skinny bodies. Such agents stop bacteria by preventing them from making cell walls. Some fungi make compounds used in drugs that are invaluable to patients undergoing organ transplants. Beside penicillin, *Penicillium* produces cyclosporine, one of the most effective and least toxic immunosuppressants known. Like the tiny bacteria and their symbiotic offspring, the protoctists, fungi embody an entire kingdom of life. They offer us a wealth of earthly delights.

The researcher R. Gordon Wasson distinguishes between "mycophilic" cultures, which love and seek out fungi, and "mycophobic" cultures, which

175

fear and despise them. This makes sense because, like many subtle and powerful substances, fungi can poison and kill as well as delight and feed. Russia is considered a mycophilic culture, and it is not uncommon for eastern and central Europeans to be well acquainted with a variety of wild mushrooms. England and America, on the other hand, are considered mycophobic countries. In anglophone countries, mushrooms generally are denigrated as toadstools and cultural ignorance of fungi is the norm. Exotic varieties of fungi are found abhorrent, dimly associated with cauldrons and the secret rites of witchcraft.

Some historians interpret references in the Indian Vedic literature to "one-footed beings that provide shade" and to the sacred drink "soma" as meaning fly agaric, a mind-altering mushroom whose species name is *Amanita muscaria*. The Eleusinian mysteries of ancient Greece probably included fungal hallucinogens, most likely the powerful fly agaric. In Siberia there is a presumably ancient rite among shamans and their followers of drinking the urine of those who have become intoxicated with *A. muscaria*. Such urine drinking rites are effective because many fungal compounds remain psychoactive after passage through the human body. All in all, we may speculate that the negative cultural attitude against fungi derived from an earlier, far more positive one, since Indo-Europeans, the parent culture of both English- and Slavic-speaking peoples, practiced Vedic rituals and imbibed soma some 7000 years ago.

Of course there are good reasons to fear fungi. These denizens of the microbial garden interfere with ordinary gardens by causing more plant diseases than do organisms from any other kingdom. All fungi produce spores, but none have undulating cell tails. Thus, fungal spores are not active swimmers like those of the funguslike protoctists. Instead, fungal spores translocate primarily through the air.

Fungal spores are carried by currents of air, landing on fruit, seed, and leaf. When moistened, the fungi spores begin to grow, quickly turning fruit soft—in less than a day in summer—spreading spots of brown, and covering peaches with green, white, or even peach-colored fuzz. Some fungi produce vomitoxins—chemicals that do just what their name suggests. Mycotoxins in rye and wheat come from the fungi naturally growing in and on these plants. It can be a fatally dangerous error to pick and eat mushrooms that you don't know. Indoors, fungi may be blamed for the black stain that creeps around the outside of bathtub caulking, wall leaks, and the liner of refrigerator doors. Have you noticed a black stain moving toward the inside of your refrigerator? Many fungi, such as fly agaric, prefer cooler climes.

Fungi are supremely terrestrial beings. They were probably the first organisms to settle the land, clearing a path for plants, which followed soon after, some 450 million years ago. Fungi have formed symbiotic partnerships with plant roots. We now know from direct study of the earliest plant fossils that plants and fungi have been evolving together since at least the Devonian period, over 350 million years ago. We know that, unlike their protoctistan counterparts, fungi never show undulipodia of any kind. This may be because, out of water, they did not need the whiplike tails to swim. But evolutionary changes ridding them of whiplashes did not entirely destroy their capacity to live in water. Today the few marine fungi that do exist also lack cell whips, suggesting strongly that ancestral fungi already were terrestrial because there

176

would be no reason to lose such a useful device for swimming if they never left the water.

Fungi extend networks of tubes containing powerful enzymes that digest food outside their bodies. As other organisms evolved on land, fungi co-evolved to digest them. Today fungi eat insect exoskeletons, hair, horn, camera lens mounting compound, film, skin, cotton, feathers, and wood—just about everything but certain plastics devised by humans. But these have been synthesized only very recently. Give them time.

▪ MATING MOLDS ▪

I f you look now, carefully, you may realize that you are already familiar with the mating molds, the zygofungi. It is highly likely that you can find the culprits in your refrigerator today. Not, of course, to imply any lack of cleanliness or housework on your part, but the mating molds have very tenacious habits. They know what they like. Nearly all the fuzzy stuff accumulating on food left too long in or out of the refrigerator contains mating molds. Simply leave some food around for a while, or gather up some suspicious-looking leftovers, and observe the more suspicious parts under a microscope.

Despite their names, the mating molds reproduce by spores and do not have to mate in order to reproduce and grow. When they do mate, or conjugate, special fungal tubes (hyphae) with "lips" at the ends grow toward each other and "kiss." No sperm or egg is formed. Upon contact the lips swell and a strange dark sporehead (zygosporangium) forms. Nuclei from both partners enter the sporehead and perform an orgiastic dance, first of fertilization and later of meiosis and spore formation. If two different mating types are needed for mating, these fungi are called "heterothallic"; if one spore grows into a thready "autosexual" fungus capable of mating with parts of itself, it is considered "homothallic." (Thallus is the word used to describe the body of an alga or fungus.) Because like all fungi mating molds produce no swimming cells, they must have other ways to effect their dissemination. A large number of fungi form windblown spores, but this is not the only mechanism.

The mating molds rapidly grow in environments rich in sugar and starch such as peaches and bread, the favorite habitat of common bread mold, *Rhizopus stolonifer* (Figure 12.1). In fact, a piece of bread may make an excellent trap for *R. stolonifer*. Just keep the mold from drying out by placing the bread in a plastic bag until you are ready to take a small dark pinch of it to be viewed under the microscope. Pick apart the blackest moist portions with a tweezer for examination.

One of the best places for the amateur naturalist to find mating molds (and other fungi, including little-known or even undiscovered varieties) is on horse dung. Although of no use to the horse, dung provides fungi with a rich source of cellulose, nitrogen, and other nutrients passed up by animals. While other fungi might thrive in dung, mating molds are usually first on the scene. They do this by the clever strategy of being eaten. But living where they do presents a problem: few animals eat their own dung. *Pilobolus crystallinus*, perhaps the most common dung-inhabiting fungus, has solved this problem in an

FIGURE 12.1

Rhizopus, the black bread mold fungus, looks like black fuzz to the unaided eye. Shown also is the sporophore with spores inside it.

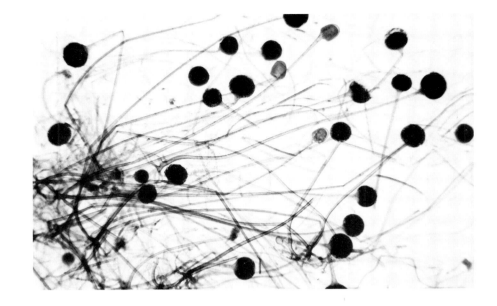

unusual and spectacular fashion. At the base of each sporehead, an unbranched structure two to four centimeters in height, is a lenslike structure called a vesicle and a light-sensitive "retina," which aims the entire sporehead toward the light. Absorption of chemicals beneath the sporehead causes mounting pressures of greater than 100 pounds per square inch to explode and fire the sporehead up to two meters from the manure—the equivalent, in a circus, of a six-foot-tall "human cannonball" being fired a distance of two football fields. By this brilliant evolutionary expedient, *Pilobilus* sporeheads fall on the grass, which is eaten by horses and other grazing animals. There is even a famous dance company called Pilobilus in honor, no doubt, of the agility of this fungal ruse.

Many other coprophilous (dung-loving) fungi have evolved similar devices to ensure they are eaten and thus end up in dung, though none are so spectacular as *Pilobolus*. *Kickxella* species, which grow on the dung of rats and bats, are thought to propagate by attaching their sticky filaments and complex coils to dung deposited along animal trails; the dung then becomes entangled in the hair of these carnivores and is spread during grooming.

Some mating molds may be useful in insect pest control. *Entomophthora muscae*, for example, infects houseflies. Infected flies can be detected by their languid manner. Full of fungi, they crawl openly on windows and die. Sporeheads, propulsively disseminated beyond the rotting external skeletons of the houseflies, leave behind whitish rings.

Still more ecologically crucial are the Endogonales, mating molds which inhabit the soil and live only in association with and generally inside of plant roots. These fungi accumulate phosphorus, a nutrient needed by all life forms but one that often limits the growth of plants, which starve without it. The roots of nearly every broad-leaved shrub and tree are intertwined with these mating molds, which, once inside, form little bulbs (vesicles) and tiny bushy branches (arbuscles). The infected roots are perfectly healthy with these mycorrhizoids, or internal fungi. Indeed, in nature neither plant nor the fungus grows alone. In fact, the oldest fossils of land plants, those of the Rhynie chert

178

in Scotland, contain microfossils of fungi entwined in the simple rootlike structures of plants. Endogonales may have evolved with the ancestors of plants from the beginning, the combination making each other stronger as they moved to dry land and transformed it into primeval woodlands.

▪ ASCOFUNGI ▪

Ascofungi (traditionally called ascomycetes but, in the five-kingdom scheme, called ascomycotes), include many of the most familiar fungi, such as brewer's and baker's yeasts, morels, truffles, and pink bread molds (Figures 12.2–12.5). They are named for their ascus, a special microscopic capsule formed as a result of the peculiar mating of these fungi. Like peas in a pod, ascofungal spores, or ascospores, form within the ascus. Most ascofungi also can reproduce in the absence of mating, in which case the hyphae directly segment to become spores. Asexual ascofungal spores are called conidiospores. When damp and well fed, conidiospores grow into new hyphae—the long, branching tubes that join to form a mycelium, or body, of the fungus. This mycelium, which can be seen with the naked eye, is a visible cottony mass made of the slender, branched fungal tubes.

Ascofungi are important to life as a whole because they break down tough plant and animal parts, such as the structural materials of wood, cellulose and lignin, as well as collagen, found in connective tissue and bones, and keratin, the basic protein of hair, horn, feather, claw, and skin. By breaking down and recycling the components of such otherwise indigestible materials, ascofungi open the way for other organisms to use the chemicals that were once locked up in these resistant materials.

But metabolizing keratin also has its bad side. The same type of organisms capable of digesting the keratin of dead animals may also be capable of

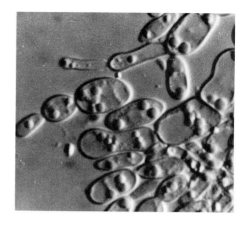

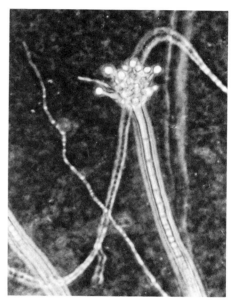

FIGURE 12.2
The ascofungus *Aspergillus* under the light microscope. Cells of the fungi are shown at the top, conidia (asexual spores) on their stalks are shown in the lower photo. (Photo courtesy of William Ormerod.)

FIGURE 12.3
Morchella (morel), an edible ascofungus. (Photo by William Ormerod.)

FIGURE 12.4
Ascomycote fungi from a book by
Ernst Haeckel (see legend to Fig.
1.2). The eight ascospores of the
ascus are shown best in the lower
middle drawing.

digesting keratin from live ones. Keratin-eating fungi cause skin maladies
such as athlete's foot, jock itch, and ringworm of the scalp. Such irritations
induce the skin cells to duplicate more often to replace dead skin. But the
flaking skin provides more keratin for the fungi and when the skin flakes it
spreads the fungus. *Candida albicans*, a microbe normally found in people,
may overgrow and cause diaper rash, yeast infections, thrush, and other
ailments whenever skin conditions become warmer, wetter, and more generally to the fungus's liking. Paradoxically, for those used to considering
bacteria only as germs, a decrease in surface populations of the bacteria called
lactobacilli actually stimulates growth of yeast associated with skin maladies.
The bacteria, which are normally present on the skin, tend to starve out the

180

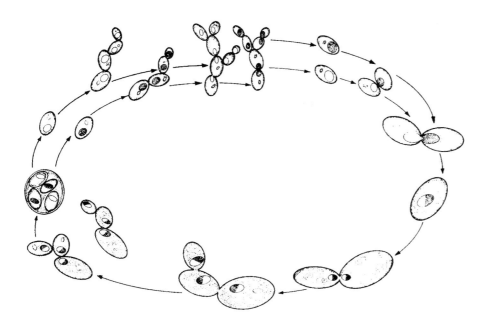

FIGURE 12.5

The life cycle of yeast *Saccharomyces cerevisiae*. Depicted is sex (fusion) and growth by budding division. (Drawing by J. Steven Alexander.)

yeast by metabolizing such carbohydrates as glycogen upon which *Candida* also thrives. If a decline occurs in the skin bacterial population, the yeast moves in and grows.

Yeasts do not produce hyphae; they are fungi with the protoctist habit of living in a free state as single cells. Most yeasts are ascofungi; the rest are basidiofungi (see next section). Some yeasts mate: the cells attract and come together, a bit of cell wall dissolves, and they conjugate or fuse to form cells with nuclei having two sets of chromosomes (Figure 12.6). After conjugation they undergo meiosis to form the characteristic ascus with four or eight ascospores.

Yeasts ferment sugars to alcohol and are used extensively in the manufacture of beers and wines. Fine wines depend on the growth of fungi normally present on the grapes' skin. Beer is made by allowing amylase, an enzyme found naturally in barley, to break down the grain's starch into sugar. Then yeasts are added, *Saccharomyces carlsbergesis* for lager and *Saccaromyces cerevisiae* for ale, to carry out fermentation. The effervescence of champagne and beer also comes from yeasts, which in breaking down sugars convert them into alcohol and bubbles of carbon dioxide gas. Bread rises from a similar process of gaseous bubbling carbon dioxide; the intoxicating ethyl alcohol of bread, however, is not ingested because it evaporates during baking.

Some ascomycetes have been researching ways to kill their bacterial competitors for hundreds of millions of years. We use penicillin, produced by *Penicillium chrysogenum*, as a near-panacea to stop the spread of bacterial infections. This biochemical works by stopping many forms of bacteria from growing their cell walls, allowing *Penicillium chrysogenum*, in turn, to spread. Although *Penicillium* molds are known for their production of antibiotics, they are also used in food production. *Penicillium roqueforti*, as its name suggests, is used extensively in blue cheeses while *Penicillium camembertii* gives brie and camembert cheese their exotic creamy textures.

The blue-green veins in roquefort cheese are the multitudinous conidiospores produced by the nuclear divisions of the mold hyphae. Outside the cheese, the same organism, *Penicillium roqueforti*, can produce PR mycotoxin, a potent fungal poison. Take a look at these cheeses and their fungi: scrape off a sliver of cheese and put it on a bit of clean bread for the fungi to grow on first. Ascofungi are among the easiest members of the microcosm for the amateur naturalist to observe.

Mysterious reports from Indochina of so-called "'yellow rain" have been traced to other fungal toxins: zearalenone and trichothecenes. These are made by *Fusarium*, a common ascofungus with spindle-shaped spores that overgrow the tissue of food plants and can lead to agricultural losses. *Fusarium*

FIGURE 12.6

Basidiomycote fungi from a work by Ernst Haeckel (see legend to Fig. 1.2).

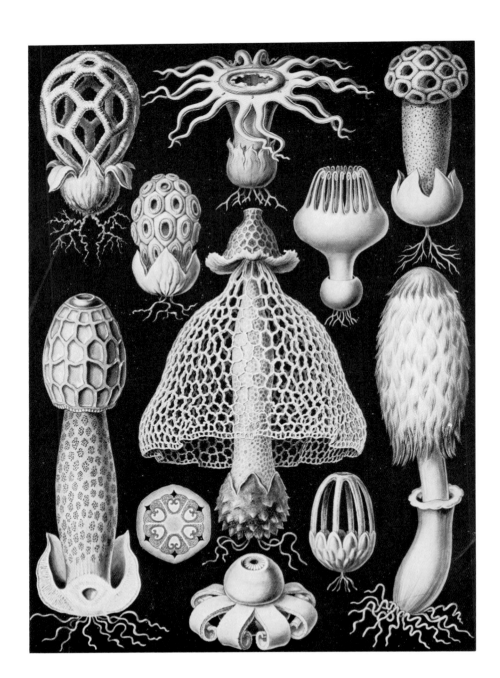

can grow under extremely low concentrations of oxygen. The toxins in question were found in Asia in the urine, blood, and tissues of victims who experienced everything from vomiting, dizziness, and diarrhea to hemorraging and death. Since these mycotoxins were restricted to areas where the yellow rain had fallen, it has been suggested that they were added to clouds as a form of biological warfare during Vietnamese invasions of Laos and Kampuchea (Cambodia) in 1975. (One interpretation is that the "yellow rain" was bee excrement, colored yellow from ingested pollen.)

The dizzying array of biochemicals produced by ascofungi extends from the horrible to the sublime. Truffles are subterranean fungi that grow symbiotically with the roots of oak and hazelnut trees, and that depend on mammals for dispersal. Once dispersed by wild boars in the French woodlands, truffles—especially the "diamond" of haute cuisine, the Perigord truffle, *Tuber melanosporum*—are now disseminated by trained dogs and pigs who, captivated by their aromas, unearth them for human consumption, disseminating their ascospores in the process. (Fascinatingly, an old story of the aphrodisiac powers of truffles may be confirmed by the recent discovery in them of alpha-androsterol, a steroid found on the breath of male pigs. Presumably this steroid, evolutionarily meant to serve as a pheromone to attract female pigs, also explains the exclusive talent of female pigs for finding truffles. This obscure steroid compound also has been detected in perspiration under the arms of men and in the urine of women. The possibility that truffles, because of the presence of minute amounts of this biologically active steroid, may be a true aphrodisiac is suggested by a study which showed that men rated pictures of clothed women more attractive after having sniffed alpha-androsterol!)

The fungi of lichen symbioses are also ascofungi. Lichens, formed by an interliving of ascofungi with either cyanobacteria or green algae, are diverse and widespread. Capable both of photosynthesis (due to the algal or cyanobacterial component) and desiccation resistance (due to the fungal component), lichens like *Parmelia* and *Cetraria* (see Figure 12.11) are found all over the world. On tree bark, rocks of the shore and even those of gravestones, lichens are quite sensitive to airborn pollutants. They generally choke on city air. Because of their sensitivity to such heavy metals as mercury and lead, lichens have been used as pollution indicators.

Another amazing group of fungi are the laboulbeniomycetes, which all parasitize insects. Some of them are so picky they will infect only one sex or even only one front limb of only one sex of the affected insect species.

▪ BASIDIOFUNGI ▪

Basidiofungi (basidiomycotes, basidiomycetes) include the puffballs, stinkhorns, smuts, jelly fungi, rusts, and common supermarket mushroom, *Agaricus*. These inhabitants of the microcosm sometimes grow big enough to emerge right into the macrocosm of everyday life (Figures 12.7, 12.8). The basidiofungi share ancestry with the ascofungi but are distinguished primarily by a reproductive structure called the basidium. The basidium, or club, originally may have evolved for ejecting spores but was modified over time. Basidiofungi reproduce sexually when fungal tubes, called hyphae (Figure 12.9), sometimes from the same spore, grow together

FIGURE 12.7

Cup fungus, *Peziza* sp., a basidiofungus. (Photo by William Ormerod.)

FIGURE 12.8

Puffball myxomycote *Lycogala epidendron*. (Photo by William Ormerod.)

and mate. The fungal tubes reproduce into the fluffy, visible network characteristic of fungi, known as the mycelium. The tubes are generally separated by dividers called "septa" and, during one stage in their lives, each segment contains not one but two nuclei. This stage, which also occurs in the ascomycetes, may last throughout the entire life cycle of basidiomycetes. Some fungal experts consider ascofungi and basidiofungi so like each other as to be merely subgroups of a single phylum of Dikaryomycotes, Greek for "double-nuclei fungi."

The jelly fungi grow on rotting wood. They have basidia shaped like tuning forks and may be found in damp wooded areas as bright yellow, gelatinous growths. Many of the gill-less fungi, or Aphyllophoriales, of which there are some 1200 species, are also wood digesters. They include the chanterelles, the horn of plenty, the dry rot fungi, the paint fungi, the bracket fungi, tooth fungi, and the club and coral fungi. *Schizophyllum commune*, a roadside mushroom, lives on dead branches. It is easily grown and mated in the laboratory, and so is a subject of genetic research.

Capped and gilled, toadstools and mushrooms are fleshy but ephemeral manifestations of fungal tubes too small to be seen. "Fairy rings" represent a subvisible advance of the growing mycelium equally in all directions from a group of spores. The spores become evident by the appearance of mushrooms and toadstools. Although they last all year long, the microscopic fungi "fruit" into visible structures only once a year, usually in late summer or fall, and then only for a brief time, ranging from hours to several weeks.

An enormous number of mushroom fungi with gills, for example the agarics—field mushrooms, including edible varieties—are symbiotic with cells of the roots of trees, explaining the proliferation of fungi in woodlands. The deadly white "destroying angel," *Amanita virosa*, is an agaric as are the other *Amanita* species, ranging from the lethal *Amanita phalloides* through the hallucinogenic *Amanita muscaria* to the tasty *Amanita caesaria*, the favorite mushroom of the Roman emperor Claudius. History has it that Claudius's scheming wife, Agrippina, who enviously desired the empire for her son, Nero, secretly spiced her husband's dish of edible mushrooms with the poisonous juice of *Amanita phalloides* (see Figure 7.0). Claudius countered her witches' brew by vomiting, but the devious Agrippina ultimately prevailed—her imperial physician poisoned Claudius by enema instead.

Both hallucinogenic and non-hallucinogenic species of *Psilocybe* are agarics, as are other "magic" mushrooms. Fungi are so varied that they seem to have been spawned from the fantasy of an imaginative writer: just look at the giant puffballs, which may grow several feet in diameter, and the earth stars, which look like breasts adorned with the leaves of a jester's cap. The order Phallales represents the stinkhorns. *Phallus* and *Mutinus* are penis-shaped stinkhorns whose stench, reminiscent of decaying meat, attracts flies. Stinkhorn spores are disseminated by the production of a sticky material that causes them to adhere to the legs of the visiting flies.

Not all basidiomycetes are repulsive. Many beside the common supermarket variety, *Agaricus brunnescens*, are edible. Delicious basidiofungi include the girolles or chantarelles of the species *Cantharellus cibarius*; the filling for crêpes à la Bordelaise, *Boletus edulis*, which is known as a *cepe* in France, a *seta* in Spain, a *Steinpilz* in Germany, and a *porcini* in Italy; the saffron milk cap, depicted on frescoes at Pompeii; the poisonous-looking blewitt, a purple or blue mushroom sold in central and northern England; the oyster mushroom; the fairy-ring mushroom, often dried and used as a spice; the honey mushroom, *Armillaria mellea*, which should be well cooked; the shiitake mushroom, *Lentinus edodes*, of oriental cuisine; the enokitake or velvet stem; the nameko; the very expensive matsutake, or pine, mushrooms of Japan; and the cloud or tree-ear mushrooms of China, which are found in many Chinese dishes and are believed to be helpful in decreasing blood clotting and heart disease. People of all cultures have interacted with basidiofungi. You can certainly find some to observe. Take the gills or pores from the underside of a mature mushroom cap and shake a few spores from them, or wait until spores form. Mycological studies make for marvelous paths from the macrocosm into the microcosm.

FIGURE 12.9
Fungal hyphae from the fungus *Aspergillus*, a common black mold.

185

FIGURE 12.10

Various species of lichens from a
work by Ernst Haeckel.

▪ LICHENS ▪

As we have seen illustrated throughout the microcosm, symbiosis is one of the salient principles of life on Earth. Those lifeforms that cooperate, help each other, and form collectives, survive. Some viciously exploit others, preying on or laying waste to their victims whose numbers decrease to the verge of extinction. But such situations don't endure. The tormented retaliate or vanish, depriving their tormenters of whatever they were used for and even of life itself. The cunning killers of the world, though they may grow briefly like weeds at the expense of those on whom they depend, often give way to the nice guys, the partners, the symbionts.

It is therefore fitting that our last fungal phylum is composed of lichens, symbiotic beings that combine members of the kingdom Monera or Protoctista with the kingdom Fungi (Figure 12.10). The cells of individuals from these kingdoms are themselves the result of bacterial symbioses. Thus bacteria have come together to make protoctists, which have diverged to form fungi, which have come together again and merged with their ancestors to form lichens (Figure 12.11).

Lichens appear as a kind of dry yellow, green, or orange "moss" that covers naked rocks. This unusual power to grow on mineral surfaces exemplifies the process by which protoctistan algae and fungi have teamed up in intricate symbioses to settle the land. Although the fungal partners of lichens are really ascomycetes or basidiomycetes, their role in the lichen partnership has transformed them to the point that they deserve their own phylum grouping. For example, lichens not only produce acids and pigments that neither of their partners can produce alone, but only together do they produce complex tissue.

FIGURE 12.11
Parmelia, this lichen is from the northeastern United States. (Photo by William Ormerod.)

187

There is not just one type of fungal partner in lichens, but rather more like 25,000. Each distinct fungal species defines a distinct species of lichen. The algae that form the other part of the lichen partnership are also somewhat diverse, though far less so than the fungi. Most commonly found as the photosynthetic partners in the fungal symbiosis making lichen are the green algae *Trebouxia* or *Pseudotrebouxia*. These walled, immotile algae are so dependent on their fungal partners that they no longer can be found growing by themselves. Many fungi with cyanobacterial, rather than algal, partners also are known: *Nostoc* the nitrogen fixer, is a filamentous blue-green that often breaks into unicells within the comfortable confines of its fungal host. Owing to such a great diversity of partners, it is thought that lichens evolved from symbioses not once but thousands of times.

The persistent lichens may grow as slowly as a couple of millimeters per century. Although they need an alternation of wet and dry periods, and are easily poisoned by air pollutants such as sulfur dioxide, lichens are indispensable to ecology. By settling on rocks, they break down solid inorganic surfaces, which leads to leaking out of phosphate and nitrate compounds into the surrounding puddles, ponds, or streams. The lichens are crucial to generating food for plant roots and for other soil microbes. For these reasons, lichens have been called agents of the weathering process. They probably have been accelerating the pace of plant, and therefore protoctist and animal, life since the Mesozoic Era when the first fossils of these symbionts appear.

FUNGI TEACHING MATERIALS

VIDEO
Life histories of the common fungi	16 min
Five kingdoms of life	10 min

SLIDES The fungi kingdom 20 (2x2) projection slides

HANDS-ON SCIENCE UNITS

What happens to trash and garbage? An introduction to the carbon cycle.

Teacher's guide
Photosort
Carbon cycle poster and guide
Life histories of the common fungi video

Five kingdoms poster
Teacher's guide to the five kingdom poster
Live cultures of fungi (e.g., *Penicillium*, *Saccharomyces*)

All available from Ward's National Science Establishment, Rochester, NY. 1-800-962-2660.

T·H·I·R·T·E·E·N

THE MACROCOSM AND BEYOND

▪ VISIBLE BIOTA ▪

The organisms discussed in this book are a sampling of the vast realm of microbes. Evolution, in the form of bacteria, protoctists, and fungi, was significantly augmented by the appearance of plants and animals in the modern Phanerozoic Eon. Prior to a billion years ago there were no plants or animals. And, indeed, it is clear that plants and animals did not pop out of thin air but evolved from the beings of the microcosm. You can see this by taking a spoon and gently scraping some cells from the inner surface of your cheek. If you look under a microscope you will see how they share the nucleus and mitochondria cellular design with fungi and protoctists. Just as protoctists have lashing tails that speed them about on the search for food or sex, so do your sperm cells or oviduct cilia display undulating tails, undulipodia whose structure is the same as that which moves genetic and other materials through the liquid medium of the microbial cell. When cut open, the undulipodia from *Chlamydomonas* or *Paramecium*, the sperm tails of brown algae or moss plants, all reveal the same arrangement of microtubules as yours. Our cells and those of microbes inside termites chomping away on a tree deep in the forest bear the unmistakable imprint of common ancestry. (The fungi have no such whips. Perhaps they lost them, or perhaps they never had them.)

Beyond our nucleated cells and those of the nucleated microorganisms lies the predominant form of life on this planet—the prokaryotes, or bacteria. Judging by their similarities to the aerobic and light-using parts of cells with nuclei, it is logical to assume that bacteria joined together in mutual dependencies to form nucleated cells. The design in Figure 1.4 shows the five kingdoms of life. Moving upward from the bottom retraces the probable development of major life forms. First, bacteria emerge from the nonliving environment. They evolve along different lines into various types, but some then come together to form larger cells in the next kingdom of cells, the protoctists. The protoctists themselves evolve myriad forms. These, as shown in the previous pages, exist plentifully to the present time. Only later do fungi,

189

plants, and animals evolve. The older kingdoms, displayed outside the shaded area, represent the majority, both numerically and historically, of life. The plants, animals, and, to a lesser extent, the fungi, represent the recent macrocosm, the biota perceived by naked human eyes. But as this book has tried to show, behind, beyond, within, and around them remains the ancient subvisible world, the realm of the microbes.

GLOSSARY

Abyss   Ocean depths; deep region below about 500 fathoms or 1000 meters.

Acidity   Capacity of water to react with hydroxide (OH^-) ions (see pH).

Aerobe   An organism that requires the presence of molecular oxygen (0_2).

Alga   Plastid-containing protoctist that performs oxygenic photosynthesis.

Alkalinity   Capacity of water to react with hydrogen (H^+) ions (see pH).

Ameboid   Shaped like an ameba; having a cell form with ever-changing cytoplasmic protrusions, or pseudopods.

Amoeba   Traditional English spelling of ameba; a free-living ameboid cell.

Anaerobe   An organism that lives in the absence of molecular oxygen (0_2).

Animal   Member of the kingdom Animalia; eukaryotes that develop from diploid blastular embryos.

Animalcule   Archaic term meaning "small animal" or moving microbe.

Antibiotic   Substance produced by one species of organism used to inhibit the growth of members of another kind.

Asexual reproduction   Any process (e.g., budding, fission) that augments the number of organisms derived from a single parental cell or organism.

Atmosphere   Gaseous mass enveloping a planetary body.

Autotrophy   Nutritional mode of organisms that can form their own macromolecules from inorganic carbon compounds (such as carbon dioxide) and obtain energy from the oxidation of inorganic compounds (e.g., ammonia, methane, hydrogen sulfide), or directly from light.

Bacteria   Monerans; prokaryotes, all organisms lacking membrane-bounded nuclei.

Banded iron formation   Distinct type of sedimentary rock consisting of alternating layers of more and less oxidized iron oxides embedded in a chert matrix; most of the economically important concentrations of iron in the world are found in Proterozoic (2500 to 570 million years ago) banded iron formations.

Binomial nomenclature   System giving each species of organism a genus (more inclusive group) and a species (less inclusive group) name.

Biosphere   The place where the sum total of all living things on Earth (the biota) resides; the environmental system of life at the surface of the Earth.

Biospherics research   Investigation of self-sustaining (closed to matter, open to energy) living systems.

Biotechnology   Science using the growth, metabolism, or chemistry of living organisms for the commercial production of food, medication, glues, or other products of human use.

Blastula   Animal embryo; a hollow sphere composed of a single layer of cells.

Bloom   Sudden large increase in a population (e.g., algae, bacteria) due to increased nutrient availability.

Carbohydrate   Organic compound consisting of a chain or ring of carbon atoms to which hydrogen and oxygen are attached in a ratio of approximately 2:1; carbohydrates include sugars, starch, glycogen, and cellulose.

Carbon dioxide CO_2, inorganic gaseous component of atmospheres and soil.

Cell wall   Generally rigid external structure produced by cells and composed primarily of cellulose and lignin in plants, chitin in fungi, and peptidoglycans (networks of amino acids and sugar molecules) in bacteria; it is absent or of various composition in protoctists.

191

Cellulose Long-chain carbohydrate that is the primary component of plant cell walls; composed of repeating sheets of linked glucose molecules.

Cenozoic Era of geologic time from 70 million years ago to the present.

Chitin Tough, resistant, nitrogen-containing long-chain carbohydrate found in exoskeletons and fungal cell walls.

Chlorophyll Green pigment responsible for absorption of visible light in photosynthetic organisms.

Chloroplast Green plastid; a membrane-bounded photosynthetic organelle containing chlorophyll.

Cholesterol Important lipid compound imparting flexibility to the membranes of eukaryotes; also the precursor of many hormones.

Cilia (singular, cilium) Short undulipodia (singular, undulipodium).

Classification The grouping of organisms into an evolutionary hierarchy; the formal hierarchy proceeds from the largest (most inclusive) to the smallest group: kingdom, phylum, class, order, family, genus, and species.

Clone Cells or organisms derived from a single cell or parent.

Colony Cells or organisms living together in permanent but loose association. Any member of a colony is capable of further growth.

Community Unit of nature comprising populations of organisms of different species living in the same place at the same time; microbial communities are those lacking significant populations of animals and plants.

Crystal Solid compound whose atoms have a regular internal repeating structure.

Cyanobacteria Oxygen-producing, photosynthetic bacteria containing chlorophyll *a* and blue-green colored pigments.

Cyclosis Streaming of cytoplasm within a eukaryotic cell.

Cyst Encapsulated and often dormant form of an organism. Propagule.

Deep sea vents Openings on the ocean floor where chemically enriched plumes of hot water are extruded because of tectonic forces.

Desiccation resistance Ability of an organism to withstand being completely dried out.

Detoxification Metabolic removal or conversion of harmful chemicals or toxins.

Diploid State of a eukaryotic cell in which there are two sets of chromosomes in the nucleus.

DNA Deoxyribonucleic acid; a long molecule that stores in nucleotide sequences the genetic information of a cell. DNA is capable of replication and directing RNA synthesis.

DNAase Enzyme that degrades DNA into its component nucleotides.

Ecosystem Unit in nature comprising communities of organisms in which the biologically important elements (carbon, nitrogen, sulfur, phosphorus, and so on) cycle completely within the unit.

Electron microscopy Technique employing a focused beam of electrons, perreitring the visualization of biological and other materials at very high magnifications (e.g., 10^5 times) (see light microscopy).

Endospore Bacterially produced desiccation- and/or heat-resistant propagule.

Embryo Early development stage of multicellular organisms; characteristic of all members of the animal and plant kingdoms.

Ergot Fungal disease of rye flowers associated with production of hallucinogenic, poisonous, and medicinal compounds.

Eukaryote Cell, or organism composed of cells, having membrane-bounded nuclei and organelles, and having DNA arranged in chromosomes; includes all protoctists, animals, plants, and fungi.

Evolution Term referring to the phenomenon and mechanism of: common ancestry, appearance, diversification, change, and extinction of organisms throughout the history of life on the Earth.

Fermentation   An anaerobic heterotrophic type of metabolism; the degradation of organic compounds in the absence of oxygen, yielding energy and organic products.

Fertilization   Fusion of two haploid cells, gametes, or gamete nuclei to form a diploid zygote.

Flagella   Long, thin, solid, extracellular bacterial motility organelles composed of the protein flagellin; not to be confused as they usually are with undulipodia.

Floc   Loose, open-structured mass formed in a suspension due to the aggregation of small particles.

Fluid   Any substance that flows, either liquid or gaseous.

Food chain   Linear sequence of organisms related to one another as producer, primary consumer (e.g., prey), and secondary consumer (e.g., predator).

Food web   Set of interactions among organisms, including producers and consumers, through which energy and materials move within a community or ecosystem. Many interacting food chains.

Fruiting body   Bacterial or protoctist structure that contains or bears cysts, spores, or other propagules.

Fungi   Chitin-walled, usually filamentous eukaryotic organisms, lacking undulipodia and embryos and that reproduce by spores (e.g., molds, mushrooms, yeasts).

Genetic engineering   Manipulation of genetic material (DNA) to alter the composition or amount of nucleic acid or proteins synthesized.

Germs   Common term for pathogenic microorganisms (e.g., some bacteria and yeasts).

Glycogen   Long-chain complex carbohydrate; the major storage substance of most animals and fungi.

Gram stain   General bacterial stain used to distinguish the two distinct types of bacterial cell walls.

Habitat   Characteristic site or local environment where an organism naturally grows.

Haploid   State of a eukaryotic cell in which there is a single set of chromosomes in the nucleus.

Heterothallic   Referring to cloned organisms that require mates; have only one type of sexual structure in a clone.

Heterotrophy   Nutritional mode of organisms that gain both carbon and energy from organic compounds.

Homothallic   Referring to cloned organisms that do not require mates; have more than one type of sexual structure in a clone.

Immune   Having resistance to disease or condition through the production of antibodies.

Immunology   Branch of science studying the immune system of organisms, i.e., the response to antigens and the production of antibodies.

Infection   Abnormal state produced by a transmissible pathogenic microbe in a plant or animal.

Inoculation   Introduction of an organism into a culture medium or environment.

Legumes   Group of flowering plants producing two-seamed pods (e.g., alfalfa, soylbeans, peas).

Light   (optical) microscopy Technique using visible light and lenses to magnify an image (magnifications 10- to 10^3 times).

Macromolecule   Long molecule, includes proteins, nucleic acids (DNA, RNAs), and carbohydrates such as starch and cellulose.

Macronucleus   Larger of the two types of nuclei in ciliates; required for asexual growth and division.

Mastigote   Undulipodiated eukaryotic protist.

Methane   Gas with the chemical formula CH_4.

Meiosis   Cell division process involving the reduction by half of chromosome numbers.

Membrane   Single or double layer of phospholipid molecules bounding an organism or organelle.

Mesozoic   Era of geologic time from about 230 to 70 million years ago.

Metabolism   Sum of all enzyme-mediated chemical conversion pathways; characteristic of all organisms.

Microbe   Microscopic organism; microorganism; includes all bacteria, all yeast and many other fungi, and many protoctists.

Microbial mat   Community of microorganisms forming a flat, cohesive structure; living precursors of stromatolites.

Microbiology   Branch of science devoted to the study of microorganisms.

Microbiotechnology   Use of microbes in industrial processes (e.g., brewing, baking, genetic engineering).

Microcosm   Subvisible world of microorganisms (bacteria, protoctists, and microscopic fungi).

Microfossil   Any remains, impressions, or trace fossils of microscopic organisms preserved in the geologic record.

Microorganism   *See* microbe.

Microtubule   Slender, hollow proteinaceous structure found in mitotic spindles, undulipodia, and other intracellular structures.

Mildew   Superficial, usually whitish, growth produced by certain fungal species.

Mineral   Naturally occuring, homogenous, crystalline element or inorganic compound.

Mitochondria   Intracellular membrane-bounded organelles in which the chemical energy of organic compounds (food molecules) is transferred to cells by respiration.

Mitosis   Cell division process involving the retention of the original set of chromosomes and thus preserving chromosome number.

Mold   Common name for a usually white fuzzy growth, usually the mycelium of a fungus.

Moneran   Member of the kingdom Monera; autotrophic and heterotrophic karyotes that include cyanobacteria and all other bacteria.

Morphology   Study of form and structure.

Multicellularity   Condition of an organism found in nature composed of more than a single cell.

Nitrogen fixation   Bacterial metabolic process involving the incorporation of inorganic atmospheric nitrogen (N_2) into organic nitrogen compounds.

Nucleus   Characteristic intracellular organelle of eukaryotes; membrane-bound structure that contains most of a cell's genetic information in the form of chromatin.

Ontogeny   Course of development of an individual organism.

Organelles   "Little organs"; distinct intracellular structures (e.g., plastids, mitochondria, nuclei, ribosomes).

Organic compound   Chemical substance containing carbon and hydrogen atoms.

Paleozoic   Era of geologic time from 570 to about 230 million years ago.

Parasite   Any organism living on or in an organism of a different species and obtaining nutrients from that organism.

Pathogen   Disease-causing microbe.

Pesticide   Chemical substance spread on or over crops that either repels or kills deleterious insects.

pH   Convenient scale for measuring the acidity or alkalinity of aqueous solutions; equal to the negative log of the hydrogen ion concentration. Pure water has a pH of 7 (neutral); solutions having a pH greater than 7 are alkaline, less than 7 are acid.

Phage   (bacterial) Virus whose host is a bacterium.

Photosynthesis   Production of organic matter from carbon dioxide and a hydrogen donor (such as hydrogen, H_2; water, H_2O; hydrogen sulfide, H_2S) using light energy captured by chlorophyll.

Phylogeny   Family tree; relationships of groups of organisms as reflected by their evolutionary history.

Plant   Member of the kingdom Plantae; multicellular eukaryotes whose cells contain plastids (i.e., chloroplasts) and that develop from non-blastular embryos (mosses, ferns, conifers, and flowering plants).

Plastid   Cytoplasmic, photosynthetic, membrane-bounded organelle of plants and protoctists.

Pollen   In seed plants, a powdery material made of spore-like cells that are male propagules; microspores containing the male genetic complement.

Population   Group of organisms belonging to the same species and living in the same place and time.

Predator   Organism that lives by killing and eating other organisms.

Prey   Organism pursued and eaten by other organisms.

Prokaryote   Bacterial cell or organism; all prokaryotes lack membrane-bounded nuclei and other membrane-bounded organelles (e.g., plastids, mitochondria).

Propagule   Any kind of reproductive particle; a useful term when one wishes to avoid specific statements about sexual or asexual reproduction, about gametes or spores, etc. Propagules are capable of survival and growth.

Prostheca   Stalk.

Protein   Macromolecule consisting of long chains of amino acid residues.

Protist   Informal name for tiny members of the kingdom Protoctista; microscopic, few- or single-celled, eukaryotic microorganisms.

Protistology   Branch of science devoted to the study of microscopic protists and related eukaryotic organisms.

Protoctist   Member of the kingdom Protoctista; eukaryotic heterotrophic and autotrophic microorganisms and their multicellular descendants exclusive of animals, plants, and fungi. Examples are diatoms, dinomastigotes, brown seaweeds and other algae, ciliates, amebas, malarial parasites, slime nets, slime molds, and many other groups.

Protozoa   Ecological term referring to any eukaryotic heterotrophic microbe that is usually motile at some stage of its life cycle.

Psychotropic effectiveness   Capacity of a substance to produce hallucinogenic effects.

Radioactivity   Spontaneous decay or breakdown of atoms accompanied by the emission of subatomic particles and electromagnetic radiation.

Replication   Molecules duplicating process involving copying from a template; not to be confused with reproduction or sex.

Reproduction   Any process that augments the number of organisms; not to be confused with sex or replication.

Respiration   Metabolic process using oxygen, nitrate, or similar inorganic compound; involves the breakdown of organic compounds and the controlled release of energy and chemically altered inorganic compounds.

RNA   Ribonucleic acid; a component of ribosomes and molecular component of the protein synthesis apparatus of a cell.

Satellite Objects, artificial or natural, that orbit planets or stars.

Seaweed Informal term describing any macroscopic, photosynthetic marine protoctist.

Sewage Refuse liquids or waste matter.

Sex Any process that recombines genes (DNA) from more than a single source in the formation of a genetically-distinct organism.

Sexual reproduction Reproduction leading to individual offspring having more than one parent; not to be confused with asexual reproduction or replication.

Sludge Semifluid, murky mass of sediment or solid matter; often associated with industrial or sewage treatment waste products.

Spirochetes Thin, corkscrew-shaped, fast-moving, heterotrophic bacteria.

Spore Any small or microscopic propagule unit containing at least one genome; often desiccation- and/or heat-resistant.

Starch Long-chain carbohydrate composed of many repeating units of glucose sugar molecules; the chief food storage substance of plants.

Stromatolite Laminated rock, usually limestone; sedimentary structure produced by the binding and/or precipitation of sediments by communities of microorganisms, principally cyanobacteria.

Subvisible Below the resolution of the human eye.

Symbiont Any organism involved in an intimate and protracted association with another organism of a different species.

Symbiosis Intimate and protracted association between two or more organisms of different species.

Taxonomy Study of the classification and naming of living things.

Undulipodia Intracellular, yet protruding eukaryotic organelles used for locomotion or feeding, consisting of nine doublet microtubules and two central microtubules composed of tubulin, dynein, and many other proteins, but not flagellin; not to be confused with flagella.

Termite Wood-eating insect, some harbor symbiotic cellulose-digesting protists and bacteria in its hindgut.

Terrarium Open man-made ecosystem containing dry land.

Terrestrial organism Organisms that complete their life cycles on land.

Toxin Proteinaceous substance produced by one organism that has specific deleterious effects on another organism's metabolism.

Vesicles Small, intracellular, membrane-bounded sacs.

Virus Noncellular infectious agent composed of a protein layer surrounding a nucleic acid core composed of either RNA or DNA.

Vitamin Any of a number of unrelated organic chemical compounds that cannot be synthesized by a particular organism and that are essential in minute quantities for normal growth and function.

Zygote Diploid nucleus or cell produced by the fusion of two haploid cells and destined to become a new organism.

ACKNOWLEDGMENTS

First of all, thanks to Moselio Schaechter and David W. Turner for their improvements of the manuscript. Douglas Zook, director of Boston University's Microcosmos: Museum of Microbial Life, led us to little-known information about the benefits of microbes; Chet Raymo, science writer from Stonehill College, suggested ways of reaching teachers, students, museumgoers, and nature lovers unfamiliar with the world of the microscopically small.

Artists Christie Lyons, Laszlo Meszoly, Kathryn H. Delisle, J. Steven Alexander, Sheila Manion-Am, Barbara Dorritie, Robert Wylie Hyde, and James Kaczman have here rendered visible the subvisible world. We are indebted to Caroline Lupfer, Wendy Ruther, Heather McKhann, Gail Fleischaker, René Fester, Rae Wallhausser, Lorraine Olendzenski, Donna Reppard, Karlene V. Schwartz, and Landi Stone for aid in manuscript preparation. David Bermudes, Gillian Cooper-Driver, Gregory Hinkle, Stjepko Golubic, Ricardo Guerrero, Frank Round, Jenny Stricker, Lewis Thomas and others provided us with everything from microbial scuttlebutt to close-up photographs of rare organisms.

We thank René Fester, Ramon Guardans, Kristine Hyon, Lorraine Olendzenski, Iyori Wada, Rae Wallhausser, and Wells Wilkinson for index preparation; they made effective use of Indexwriter by Zachary Margulis. Financial support from the Richard A. Lounsbery Foundation, the NASA Life Sciences office, the Boston University Graduate School, and the College of Natural Sciences and Mathematics of the University of Massachusetts, Amherst are gratefully acknowledged.

Dorion Sagan
Lynn Margulis

INDEX

Italic page numbers refer to illustrations. Genera and species are italicized, the genus name capitalized.

219

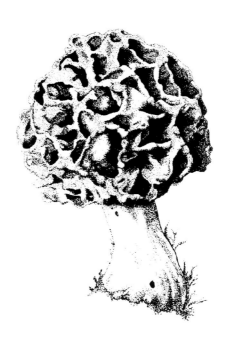